S0-AFW-457

General Physics I-II

Laboratory Manual

PHYS 2014-16

For use at

University of Minnesota Duluth

Kendall Hunt
publishing company

Cover image provided by Author

Kendall Hunt
publishing company

www.kendallhunt.com
Send all inquiries to:
4050 Westmark Drive
Dubuque, IA 52004-1840

Copyright © 2013 by Jonathon Maps

ISBN 978-1-4652-2321-0

Kendall Hunt Publishing Company has the exclusive rights to reproduce this work,
to prepare derivative works from this work, to publicly distribute this work,
to publicly perform this work and to publicly display this work.

All rights reserved. No part of this publication may be reproduced,
stored in a retrieval system, or transmitted, in any form or by any
means, electronic, mechanical, photocopying, recording, or otherwise,
without the prior written permission of the copyright owner.

Printed in the United States of America
10 9 8 7 6 5 4 3 2

Contents

iv

Rev. May 24, 2013
©Jonathan Maps

Laboratory information

The test of all knowledge is experiment. Experiment is the *sole judge* of scientific "truth."

Richard Feynman

Lab schedule and procedures

Lab meets weekly. You are expected to prepare for lab by reading through the lab in advance of your section and completing any written pre-lab exercises in the lab manual.

When you come to lab bring:

- the lab instructions and the completed pre-lab exercise;

- a calculator;

- a USB drive for storing any data collected via computer;

- a lab notebook.

    At the first laboratory meeting you must have a suitable notebook,

    - Sewn binding or strong glued binding—*not* spiral or loose-leaf;
    - Quadrille rulings (graph paper grid—*not* lined for composition);
    - Size of 8 x 10 inches or less.

You will typically work in groups of two or three on the same apparatus. You cooperate in designing procedures and taking data. Each team member must make a record of all data in a laboratory notebook as the experiment is conducted. The data pages of the notebook must be initialed and dated by the lab instructor before you leave the laboratory. You are expected to carry out basic analysis before leaving lab to ensure all necessary data was recorded and the data collected is correct. Final analysis will be done outside lab.

Please leave your lab table clean and equipment neatly organized at the conclusion of each lab period for the next group. Habitual slovenliness will lower your lab grade.

The labs may be performed in a sequence different than the order in the lab manual. Your lab instructor can tell you the next lab to be performed each week. Links to the schedule of labs can be found on the WWW at http://physics.d.umn.edu/ along with other lab information.

Changes to information in this manual will be announced in lab. Additional information on methods is contained in appendices. Please read them as they become appropriate.

Laboratory notebooks

Notebook format requirements. Your performance in the laboratory is evaluated chiefly by the written records you keep.

- Leave the first page for use as a table of contents. Keep it up-to-date. Number all pages consecutively.

- Begin each experiment entry with a brief title and remarks on the objective of the experiment. Date all entries. Give names of your laboratory partners at the beginning of each new lab.

- Initially write only on the right hand page; leave the left hand page for attaching graphs, extra calculations, after-thoughts, etc. Write in pen. If you make an error, simply cross it out. The crossing out should not make the entry impossible to read. You might find the entry useful at a later time. Don't make notes or calculations on loose sheets of paper unless specifically requested; put the information in the notebook directly.

- The entries for each experiment must include:

  - simple sketch of the experiment that shows important quantities;
    A carefully drawn and labeled sketch can convey much about the experiment and what was measured. Small, hastily drawn sketches are usually worthless. Sketches need not be great works of art to be informative.

  - description of any methods developed to perform the lab;
    Someone not familiar with the experiment should be able to understand in broad terms what was done based only on your notebook.

  - data;
    Put data in tables whenever possible. Always include units. All data should be in the notebook. When data is collected by computer, printed tables should be taped—*not stapled*—into the notebook. If

a computer-collected data table extends to many tens or hundreds of points, graphical presentation of the data is preferred. Lab notebooks must be initialed and dated by the instructor before you leave the lab.

– calculations;

Show calculations in sufficient detail so another person can follow and check your work. Always begin a calculation by showing the formula being used, then substitute in numerical values <u>with units</u>, and finally give the final numerical result with units. The following example from the density of solids experiment illustrates the required format.

$$\rho = \frac{4M}{\pi d^2 L} = \frac{4\,(52.3\,\text{g})}{\pi (1.254\ \text{cm})^2\,(5.12\ \text{cm})} = 8.27\,\text{g/cm}^3$$

If a calculation is repeated several times, show one sample in detail and incorporate the results into an additional column in your data table.

– graphs;

When you include a computer-prepared graph or a separate sheet of graph paper, tape it into your notebook—don't staple it in. Place it in a way that makes it easy for a reader (e.g. the grader!) to view it when reading your notebook. Tape it to a left hand page. It should unfold for easy viewing. The characteristics of good graphs are discussed in an appendix.

– a concise, technical summary of results and answers to any questions in the printed laboratory instructions.

Lab grading

All work required for the experiment must be kept in the laboratory notebook. A separately prepared lab report is not required. A few days after the laboratory class meeting the completed notebook is turned in to one of the locked boxes on the second floor of MWAH. The day and time labs are due will be determined and announced by the lab instructor. The graded notebooks will be returned at the start of the next laboratory meeting. See the syllabus for additional details on grading and lab policies.

In science there is only physics. All the rest is stamp-collecting.

Ernest Rutherford

1—One-dimensional motion

Object

This lab is intended to strengthen your ability to construct and interpret graphs representing motion and your understanding of the relationships between plots of position, velocity, and acceleration as functions of time. It will also provide experience with data acquisition and analysis with computer-based sensors and software.

Background

In this lab your lab team will observe a glider released from rest on a tilted air-track and construct graphs of position and velocity as functions of time, $x(t)$ and $v(t)$. You will do this first by simple observation and making some qualitative sketches. Later, more quantitative measurements will be made with an ultrasonic range finder/motion detector. This device emits a burst of high frequency sound (ultrasound) that reflects from nearby objects and the device then detects the arrival of the echo—the same idea as SONAR. The distance to the reflecting object is determined from the time it takes the sound pulse to travel out and back. The graphs obtained with the motion detector will be compared with the simple qualitative observations.

A graph of $x(t)$ can provide insight into the velocity as a function of time, $v(t)$. The instantaneous velocity $v = \frac{dx}{dt}$ corresponds to the slope of a line just tangent to the $x(t)$ curve at any point. Similarly, acceleration can be found from the slope of the $v(t)$ graph, $a = \frac{dv}{dt}$. The acceleration is also related back to $x(t)$ as a second derivative, $a = \frac{d^2x}{dt^2}$. Recall from calculus that you can relate the second derivative of a function to the curvature of that function: a function whose graph is "concave up" has a positive second derivative at that point.

The air-track provides a low friction surface for gliders to move freely along. Air forced out through small holes along the track levitates a light-weight metal glider, which can move smoothly along the track with little resistance. For best results, the air supply should be turned on and the system allowed to warm up before making measurements. The air-track expands slightly as the temperature increases. Once you begin your measurements, you should leave the air supply on until all needed measurements are completed. A measuring tape is mounted along the side of the air-track to allow rough measurements.

With the motion detector, ultrasound pulses will reflect from a small metal tab mounted on the glider. The motion detector is controlled by computer and can emit many pulses each second to produce a set of measurements of the glider's position, x, as a function of time t. The data collected by the motion detector are a series of points (t_1, x_1), (t_2, x_2), (t_3, x_3), (t_4, x_4)…. If the (t, x) points are collected frequently enough, they can be used to make a pretty good estimate of the instantaneous velocity. For example, from these data the velocity at t_3 can be estimated as

$$v_3 \approx \frac{1}{2}\left[\left(\frac{x_3 - x_2}{t_3 - t_2}\right) + \left(\frac{x_4 - x_3}{t_4 - t_3}\right)\right],$$

which is the average of the slopes of the two line segments connecting the point (t_3, x_3) to its nearest neighbors (t_2, x_2) and (t_4, x_4).

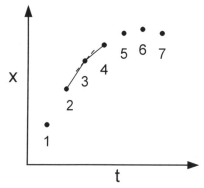

The smaller the time interval between successive measurements, the more nearly this approximation approaches the definition of the derivative as a limit:

$$v = \frac{dx}{dt} = \lim_{\Delta t \to 0} \frac{\Delta x}{\Delta t}.$$

Software that collects data from the motion detector can carry out this numerical approximation to the derivative $v = dx/dt$ automatically.

PLEASE HANDLE THE AIR-TRACK GLIDERS VERY CAREFULLY!
LEAVE AIR SUPPLY ON CONTINUOUSLY THROUGHOUT
MEASUREMENTS.
DO NOT TURN OFF AIR SUPPLY WHILE GLIDER IS IN MOTION.

Procedure

Observing and sketching graphs describing motion. Watch an air-track glider traveling on a tilted air-track. Release the glider from a point about 1 to 1.2 meter from the lower end of the track. Watch the motion of the glider through four bounces and rebounds up the track from the bumper. You can make several trials; use a consistent starting point.

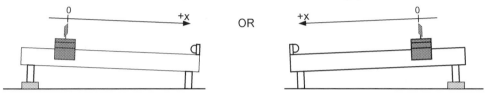

First qualitative graph. Make a *careful* qualitative sketch of the glider's position as a function of time, $x(t)$, in your lab notebook. Take the glider's starting position to define $x = 0$ and let the $+x$ direction be downhill on the track. By "careful" we mean sketching a graph that, while not precisely numerically to scale, still reflects as much about the motion as you can observe from a few repeated trials. *Don't* try to make a detailed table of data; instead use the following to guide your graphing:

- Begin by drawing axes for your graph with a ruler on a *new* page in your lab notebook: a horizontal axis for t, and a vertical axis for x. Label your axes. Leave room for another similarly sized graph directly below it on the same page.

- The release point serves as the origin of your x-axis, so your curve must begin at the graph's origin: the glider starts from $x = 0$ at $t = 0$.

- What was the greatest distance the glider reaches from its starting point? To make a rough scale for your vertical (x) axis, let 1 box of the grid in your lab notebook represent 10 cm.

- When sketching your graph of $x(t)$, take the time to make what you think are straight-line segments be straight lines—use a ruler to help. If you think some part of the graph curves, does it curve up or curve down? Draw it so.

- Does the glider rebound to the same position after each successive bounce? Make your sketch show these details as best as your observations permit.

Label points along the time axis corresponding to when each of the four bounces from the spring bumper occurs with "B_1", "B_2", "B_3", and "B_4." Also mark the times corresponding to when the glider returns closest to its starting point on each rebound by "R_1", "R_2", "R_3", "R_4."

Very roughly estimate how many seconds elapse from release to the glider's first return closest to the starting point. (Counting seconds out loud as 'one-thousand one, one-thousand two, ...' is good enough!) Note this time on your graph next to the point you labeled "R_1."

Second qualitative graph. Next sketch a qualitative graph of the glider's velocity throughout the motion. Draw new axes for this graph directly below the $x(t)$ graph. The horizontal (time) axis should use the same scale as your $x(t)$ graph. That means significant events during the motion (e.g., bounces and closest returns, B's and R's) in the two graphs should be lined up to make comparison of $x(t)$ and $v(t)$ at the same instant easy. Sketching this graph may be trickier based on simple qualitative observation, particularly when trying to decide whether segments of the graph should be straight lines or curves. Do the best you can and consider qualitative clues that are most easily available, such as: "When is the glider moving fast or slow?" and "What is the sign of $v(t)$?" when forming your graph.

Can you think of a simple way to make a very rough estimate of the velocity scale for your graph? If so, explain how and indicate that scale for the vertical (v) axis of your $v(t)$ graph.

Measurements with a motion detector. Once your sketches are complete and satisfactory—they don't have to be correct, but your lab instructor may ask you to think harder about your graphs before proceeding—you'll get an ultrasonic range finder/motion detector. The motion detector cannot measure objects too close to it. Position the motion detector about 50 cm behind your starting point.

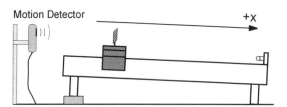

Data is collected from the motion detector using the program *Logger Pro.* A pre-configured experiment file, '1Dmotion.cmbl,' will be in the course folder on the computer desktop. Opening this file will launch *Logger Pro* automatically. Read any instructions displayed when the file is opened.

It may be easier to become familiar with the operation of the motion detector and software by initially using yourself as the target for a few minutes and recording your own motion.

Alignment of the motion detector. Be sure to align the motion detector with the air-track so that it follows the glider's motion successfully throughout the full range of motion. Aim it carefully along the air-track at the glider. Aiming slightly high can avoid spurious ultrasonic reflections from the air-track itself. If you look from the far end of the air-track toward the motion detector, you should be able to see the reflection of the top apex of the air-track in the gold-colored membrane in the motion detector. The reflection should continue in a straight line with the actual air-track, not bend away at an angle with the air-track. Keep the sound path clear of extraneous objects (books, equipment, people, computers), which can cause false echoes.

A properly aligned motion detector and a table free of obstructions will yield clean and reproducible data. It may take some care and patience to get useful data.

Use the motion detector to record the motion of the glider on the air-track through four bounces and rebounds from the bumper. Start data collection but wait 1 to 2 seconds before releasing the glider; the motion detector must be initialized each time and it does not begin collecting data immediately. You can hear clicks from the motion detector as it emits bursts of ultrasound. Make several trials until you are certain you get good, reproducible results, and have good graphs of $x(t)$ and $v(t)$.

Interpreting the graphs. Using data from a successful trial, print out copies of the two graphs for inclusion in your lab notebook and further analysis. Print each graph separately, but before printing each graph make sure you have provided correct labels for your axes, including units, and a graph title. You can later fold it in half and tape it (please don't staple!) onto the left-hand page of your notebook so that it unfolds and is easy to read for anyone reading your notebook.

Label the points "B_1," 'B_2," etc. again on your new graphs. Compare the resulting $x(t)$ and $v(t)$ graphs to your qualitative sketches. If there are qualitative differences between your sketches and the motion detector results, describe them. What are the similarities?

Use the graphs obtained from the motion detector to answer the following questions in your lab notebook. Use complete sentences and be as quantitative as possible. (By "bounce" we mean the short time interval during which the glider is in contact with the elastic bumper.) The software has a variety of analysis tools to help you answer some of these questions.

1. From the $x(t)$ graph, how does the time between the first and second bounce compare to the time between the second and third bounce?

2. How close to the original starting point does the glider come on each of its rebounds after the first and second bounces?

3. Where during the glider's motion is its velocity zero?

4. Compare the glider's velocity immediately after the second and third bounces.

5. Using the $v(t)$ graph, is the glider's acceleration ever positive? Negative? Describe when in the motion each occurs and explain clearly how you determine this from the $v(t)$ graph.

6. Calculate the acceleration of the glider while it is traveling along the air-track between the first and second bounces from your printed $v(t)$ graph. (Draw the best straight line through the corresponding portion of the $v(t)$ graph and extend this line to the graph edges. Recall that the average acceleration is $a_{av} = \frac{\Delta v}{\Delta t}$ or the instantaneous acceleration $a = \frac{dv}{dt}$. Find the acceleration by calculating the slope of this line based on the points where it intersects the edges of your graph. Show all the steps in your calculation. While we will make of software tools to fit lines and find slopes in the future, do it *by hand* for this case.

7. Without detailed calculation, compare the acceleration of the glider while the glider travels along the air-track between second and third bounces to the acceleration just found between first and second bounces.

8. What is the glider's acceleration at the top of each rebound back near the origin, when the glider changes direction?

9. Estimate from your $v(t)$ graph the average acceleration *during* the first bounce, i.e., during the short time while the glider is in contact with the bumper. $a_{av} = \frac{\Delta v}{\Delta t}$.

10. Hand-sketch a graph in your notebook of the glider's acceleration, $a(t)$, based on the motion detector $v(t)$ graph and your answers to questions above.

   Save your data and take it with you for later use. Do not expect any data you save on the lab computer to be preserved after your lab section.

   Before you leave: Return the motion detector to your lab instructor when your measurements are complete. Make sure your lab table is ready for the next group. Leave equipment neatly organized for the next lab. Please don't leave random scraps of paper, calculators, water-bottles, etc. behind. Use recycling bins down the hall, not the trash, for recyclable materials.

Logger Pro tips

The motion detector should be connected to a DIG/SONIC port of the Lab Pro or LabQuest interface, which in turn is connected to one of the computer's USB ports.

Data collection is started with the green "Collect" button on the menu bar (or from the keyboard by pressing the space bar). Data will be collected for a pre-determined length of time and (usually) plotted as it is collected. You can interrupt data collection with the "Stop" menu bar button that appears when data acquisition is in progress or with the space bar.

To print a graph, click on the graph to select it. Select landscape printing ("File – Page Setup" and select "Landscape"). Then from the menu bar select "File – Print Graph."

Logger Pro has a tool to calculate slopes over segments of the data; while you can use this to verify your results in this lab if you know how, you must still do the calculation by hand based on the printed $v(t)$ graph yourself. In later labs we will make much use of the fitting/regression tool.

A useful exercise might be this *post-lab analysis:* Discover how to use *Logger Pro* to calculate $a(t)$ from the data already collected and saved. Create another data column (from the menu bar select "Data – New calculated column." A new dialog window will open with tabbed pages: Pick 'Options' and and enter appropriate entries on the Options page. On the 'Definition' tab page you'll need to figure out how to enter a formula into the Equation box to calculate acceleration. There are several ways you might do this. The functions available in *Logger Pro* can be selected from a drop-down Function list; you can also pick out variable names, which must be enclosed in double quotes when used in the Equation box, to complete the formula. You can look at the definition of the velocity column to see an example. Compare the acceleration produced by *Logger Pro* with your rough sketch. Discuss the differences and similarities.

2—Graphical analysis of free-fall motion

Object

This lab provides experience in using graphical techniques to confirm or reject various models for falling bodies, and to analyze the graphs to determine the acceleration due to gravity and the initial velocity of a falling object.

Background

In this lab you are provided with some raw data recording the position of a freely falling object at evenly spaced instants in time. The data are in the form of small dots burned into a long strip of waxed paper. Your task is to produce a listing of time, t, and position, y, values and then analyze these data to test several possible functions, $y(t)$, that describe different kinds of motion. The analysis of each model is done by figuring out a way to plot the measured data, (t, y), or other sets of coordinates based on your measured t and y values so as to yield a straight line if and only if the model is correct. This method of analysis is often useful in determining which model or theory best describes experimental results, since it is often much easier to judge whether data accurately follows a straight line compared to following various curves. Graphical analysis of data in this way will be used throughout the physics sequence. In this lab you can use software to reduce the tedium of repetitive calculations and to prepare suitable graphs.

You will test the experimental data against three mathematical models for the motion of a freely falling object:

(1) Constant velocity motion: $y = y_o + v_o t$,

(2) Constant acceleration motion: $y = y_o + v_o t + \frac{1}{2}at^2$,

(3) Variable acceleration motion: $y = y_o + v_o t + Ct^3$.

In each case, some as yet unknown parameters are involved. For all three models, v_o is the object's initial velocity (v at $t = 0$) and y_o is the initial

15

position, $y(0)$. Models 2 and 3 also have another unknown parameter: a, the acceleration in Model 2, and C, which characterizes a time-dependent acceleration in Model 3. Graphical analysis will reveal which model is best and what the corresponding values of these parameters are for the correct model. The constant y_o in each model can be eliminated by choosing an origin for measuring position at the first dot on the tape and assigning it the coordinates $(t, y) = (0, 0)$. This simplifies the model equations a little:

$$(1) \ y = v_o t, \qquad\qquad (2) \ y = v_o t + \tfrac{1}{2} a t^2, \qquad\qquad (3) \ y = v_o t + C t^3.$$

Linearization of models. The idea here is to invent ways to re-write each model equation in the standard form for a straight line: $Y = mX + b$, where m is the slope and b is the Y-intercept of the line. The Y and X can be 'compound variables'—constructed from combinations of the raw (t, y) data. Plotting the corresponding (X, Y) points calculated from the raw (t, y) data will yield a straight line *if* the model is correct. The data will curve if the model doesn't apply to the data. This may be less abstract with a couple of examples.

As a first concrete example, consider Model 1 and its equation $y = v_o t$. Re-writing this in the standard form is trivial—it already *is* in the form of a linear equation! You can take the individual measurements y as Y and t as X so that the equation for model 1 becomes $Y = v_o X$—a straight line with slope v_o passing through the origin. The invocation of 'compound variables' is an unnecessary (and probably confusing) step for this simple case.

As a less trivial example, consider a special case of Model 2, with $v_o = 0$: an object known to be at rest at $t = 0$. The model equation is then just $y = \tfrac{1}{2} a t^2$. One possible choice of compound variables X and Y is to simply call $Y = y$ and $X = t^2$. With the substitutions $y \to Y$, $t^2 \to X$, the model equation becomes $Y = \tfrac{1}{2} a X$, which is the equation of a straight line with slope $\tfrac{a}{2}$ passing through the origin. To test this you would need to calculate t^2 values for each original point, then plot on a new graph $Y(= y)$ as a function of these new $X(= t^2)$ values, *i.e.*, (t^2, y) points. *If the data fell along a straight line when plotted this way, the model would be a good description of the data and measuring the slope would allow us to determine a.* (Note these choices for Y and X do *not* convert the full-blown Model 2 into a linear equation, since with the $v_o t$ term, these choices for X and Y turn that equation into $Y = \tfrac{1}{2} a X + v_o \sqrt{X}$, definitely not the equation of a straight line. So it is *not* an appropriate choice for testing Model 2 later in this lab!)

In seeking good compound variables, keep in mind the following:

- The compound variables, X and Y, can be combinations of only t and y. Since you only have the raw t and y measurements to calculate with, your X and Y must use only these raw data.

- A straight line is characterized by two parameters: the slope and the y-intercept. Our models also involve two parameters e.g., v_o and a for Model 2, so the parameters in the model equations must somehow appear in the slope and intercept of the standard linear equation. What you seek to do is turn each model equation separately into something that looks like $Y = (\text{constant})\, X + (\text{another constant})$.

- You can "massage" the model equation using any legal algebraic operation applied to both sides of your starting equation. The right-hand side of your manipulated equation should contain the unknown parameters (e.g., v_o and a for Model 2, and v_o and C for Model 3) and something you will call X that can involve t and perhaps y.

- Compound variables can involve combinations of y and t that are products or quotients, e.g., $Y = y/t^2$, y/t or yt might be legitimate choices if they naturally appear as a result of algebraic massaging of the model equations. Note that a combination of y and t that use use direct addition or subtraction, for example, $Y = y + t$, is not valid since these are dimensionally different quantities. The sum of meters and seconds is meaningless. Besides, it's not clear what legal algebraic operations would lead you to such a combination from the model equations (2) or (3).

Procedure

Gathering the raw data. The waxed paper tape can be taped flat to the table for measurement. Record in a table in your lab notebook the t and y values for every third dot. Since the dots are produced every $\frac{1}{60}$ second, the time interval between every third dot is $\frac{3}{60}\,\text{s} = .050\,\text{s}$.

The meter stick should be placed on its narrow edge so the rulings extend to the tape for easy, accurate measurement. The value of y_o is arbitrary in the original model equations since it depends on where you place the origin of your y-axis and which dot you select to be $t = 0$. The most convenient choice is to select the one of the top-most dots, (call this the 'zeroth' dot) as the origin: $(t_o, y_o) = (0, 0)$. Please realize that other dots have been torn off when the tape was prepared, so this dot does *not* mark the instant when the free-fall began. The object was already falling with some velocity v_o at this dot.

Occasionally a spark fails to form at the proper time and a dot will be missing. You can usually notice when a dot is missing or clearly out of place from a quick inspection of the tape. One or two missing dots on a tape are usually not a problem. Just pick a dot near the start of the tape as the origin so that the missing or questionable dot is one of the dots skipped over in counting out every third dot for measurement.

Testing Model 1. Test this model by making the corresponding graph of $(X, Y) = (t, y)$ values as suggested in the Background section. This can be done with the assistance of software. *Logger Pro* provides a simple spreadsheet-like interface for entering the raw data in tables and graphing it. The file 'freefall.cmbl' provides a template to help you get started. *Logger Pro* also provides the ability to calculate additional columns based on the raw data you enter. This feature will be useful in testing models 2 and 3.

Enter your raw t, y data in the designated columns. Examine the resulting graph, titled 'Model 1,' of $y(t)$. Make sure that the limits of your graph include the origin (0,0). You can change the upper and lower limits on the axes ranges by clicking on their values in the graph. Print out a copy of the graph for your notebook. Make sure your graphs conform to requirements *before* printing them.

- Print in landscape mode ('File – Page Setup'). You will want to tape— do not staple!—the graph into your notebook at the appropriate place when finishing your notebook for submission. This way it will unfold nicely for viewing.

- Print each graph alone ('File – Print Graph', not a screen dump) so your graph occupies as much of the page as possible.

- You should be sure to enter accurate labels for the both the horizontal and vertical axes with units, and include an informative graph title. Do not accept any default labels, titles or numerical ranges for the axes without ensuring they are complete and appropriate for your graph.

- If you later discover an error or omission in labeling or titles after printing, just neatly write in the correction by hand. Major blunders may require re-printing, but not minor things easily fixed by hand.

Do the data fall along a nearly straight line passing through the origin? If so, you can conclude equation (1) describes the motion and that model is correct. If and *only if* this is true, draw with a ruler the best straight line on your copy of the graph, and find the slope. *If the data clearly curve, the equation does not fit the data and the model should be discarded. No straight line should be drawn and no slope or intercept should be found. They would be meaningless.*

Without a straight line here, it's then time to consider the next proposed model. But with this much work done, it's always a good idea to save your data to a file, preferably on your own USB storage device. Saving early and often can prevent loss of data and the need to repeat time-consuming measurements.

Analysis of models 2 and 3 requires some creative thought on your part as described in the Background section. You need to invent ways to re-write the equations for (2) and (3) so that they look like the standard form of a straight line equation. To do this you need to identify 'compound variables'— allowed combinations of the raw measured t and y that you can assign to X and Y— to make the result look like the standard linear equation $Y = mX + b$. You'll need to fiddle around with the equations algebraically to "linearize" them.

Testing Model 2. Discuss among your lab group possible ways to make equation (2) look like the equation of a straight line and what the corresponding compound variables are. Document in your lab notebook how you arrive at your choice of compound variables and point out what should correspond to the slope and the Y-intercept.

After your lab group has decided on a set of compound variables, X_2, Y_2, that are good for testing model 2, you can have *Logger Pro* do the calculations and graphing. The template file contains a separate page with columns for X_2 and Y_2 named "Model 2 X" and "Model 2 Y." Initially these are defined to be $X_2 = t$ and $Y_2 = y$. You will need to change these definitions by double-clicking on the chosen column, or by right-clicking on the column and selecting 'Column Options' and the corresponding column, or from the menubar select 'Data – Column Options' and then the column name. A window appears with tabbed pages. On the 'Definition' page you construct the definition of your compound variable in the equation box, adjust the column title, and specify the units. The 'Options' page allows you to specify the appropriate number of significant figures to display in the table. Pay attention to these details when making changes. (A more detailed description of making these changes is at the end of this lab if you have difficulties.)

Does the graph you construct for model 2 fall close to a straight line? If so, how are the slope and intercept related to v_o and a? Adjust the ranges on your graph axes so you are sure to include the origin $0, 0$ and the range extends slightly beyond any individual data point.

Print out the graph and draw the best straight line through the data.

Extend the line you draw to the boundaries of the graph. Find the coordinates of the two points where your line intersects the graph's edges. You may find it necessary to use a ruler to determine the scale for the graph to obtain more precise values of the coordinates.

Read off the Y-axis intercept. Calculate the slope of your line based on these two points—do *not* use any individual data points, no matter how close you feel they fall to the line. Clearly show all steps of your calculation in the notebook. Include units throughout all steps of calculations. Read off the Y-axis intercept.

From these, report out your values for v_o and a clearly in you notebook. Be sure to include your printed graph in your notebook. Are these reasonable values? Is your value of a consistent with the accepted value for the accepted value of the local gravitational acceleration, $g = 9.80 \, \mathrm{m/s^2}$?

Testing Model 3. Find compound variables that linearize the equation for Model 3 and graphically test the agreement between the data and Model 3. Show the algebraic steps to document how you arrive at your choice of compound variables. Use *Logger Pro* to calculate and plot them. (The template file has a third page for this model.)

Does the graph you construct for Model 3 form a straight line? If (*and only if*) the data now fall along a straight line, find values from your graph for the slope and intercept and relate them back to the unknown parameters of the model (v_o and C). Do not attempt to find values if the data do not conform to the model! Keep in mind that at most *one* of the models can best describe the motion of the falling body and produce a straight line when properly linearized. Drawing or fitting a straight line through data that clearly curve is generally a foolish and incorrect thing to do. Only one model should be fitted to a straight line in this lab.

Computer fitting of data to a straight line. *Logger Pro* has tools for computing a best-fit line (a least squares fit or linear regression) through data (and to other equations more complicated that linear relationships). Return to your graph created for Model 2. Use the mouse to select the data in the graph and use the Linear Fit button on the menu-bar to have Logger Pro perform the fitting operation. Record the results in your notebook. What values does this fitting process give for g and for v_o?

Before you leave: Save your completed Logger Pro file with all the analysis and take it with you. Make sure all lab team members have a copy. Clean up your lab table before you leave neatly arranged for the next lab section. Don't leave scraps of paper, used tape, etc., behind.

Post-lab analysis: Calculate the percentage difference between your determined value for g and the accepted local value of 9.80 m/s^2. (Percentage Difference = $\frac{expt\_result-accepted}{accepted} \times 100\%$).

Do this for the computer-based best-fit line, too.

How many "significant figures" do you think are warranted in reporting your by-hand result for g? For the best-fit result from software? Explain how you decide this. Note: Computers and software are usually quite happy to automatically fit any data to any equation and print out a result, often with lots of digits, without regard to whether any of it makes any sense. It is up the the user (YOU!) to decide when fitting data to a particular equation is a meaningful thing to do or just a silly waste of time. It's also the user's responsibility to decide how precise the results spit back by a machine are and to report an appropriate number of digits that takes into account the precision of the original data.

Notes on good graphs.

Graphs should be large, occupying as much of a page as possible. The plotted data should generally fill as much of the graph as possible. You may need to adjust the ranges for the X and Y axes to do this. At the same time, choose ranges that put the major divisions or tic-marks at convenient, round values. Always provide an informative and specific title for every graph. Label the axes with the quantity plotted and be sure to include units. See the appendix on graphing data for more information.

Notes on using *Logger Pro* to create, modify, and plot new columns.

As an example of how to modify columns calculated by *Logger Pro*, let's assume you want to use $X_2 = t^2$ and $Y_2 = y\,t$—*not* really a good choice for model (2) since it does not linearize Eq. (2). To enter the definition for X_2, double-click on the Model 2 X column in the data table. Change the column name to X_2=t^2, and short name to t^2. Construct the column definition in the equation box. To calculate new columns based on data in other columns, *Logger Pro* requires you must put the column names in the formula and enclose them in double-quotes. To calculate t^2 you can enter "t" * "t" or equivalently "t"^2. Put in the correct units in the units box. From the Options page choose an appropriate number of significant figures to display in the data table. Click OK when you have made all the necessary changes to the column. To enter your definition of $Y_2 = y\,t$ follow a similar procedure:

> double-click on the Model 2 Y column
>
> change the column name to `Y_2=yt`, short name to `yt`
>
> enter the definition in the equation box `"y" * "t"`
>
> enter the correct units and significant figures.

You can also easily access column names and automatically insert them (with quotes) in the equation box by selecting them from the drop-down list presented in the 'Variables (Columns)' listbox. You can add your own new calculated columns to a data table from the menubar by selecting 'Data – New Calculated Column' and entering the information as above.

To change the column of data used in a graph simply click on the label for the axis on the graph you want to change and select the new column of data that should be used for as that axis.

Analysis of video data. Free-fall data may be provided in the form of a movie of a falling object. Logger Pro allows you to analyze the video frame by frame, recording the position of the object and generating a table of position vs. time data. To use data from a video, first add the video to the Logger Pro file (Insert – Movie, then navigate to the video file). You can play the movie, advance to the interesting region, and advance frame-by-frame. Enable the video analysis tools with a button at the lower right of the video display. This provides a set of buttons on the right of the video frame. The top crosshairs button allows you to record the position of an object frame by frame and populate the data table with data. Before doing so, you should do two things: (1) establish an origin—a point in the video that will correspond to $(x, y) = (0, 0)$ using the set origin tool; (2) set the scale using an identifiable reference object of known length (e.g., a meter stick in the video, by selecting the set scale button, holding the mouse button as you drag it from one end of the reference object to the other and then entering the known length of the reference.

Then, the crosshairs tool can be used to mark the position of the object frame by frame by clicking on its location. Logger Pro will advance automatically to the next frame after clicking on the object. For vertical freefall, we don't care about the x-coordinate.

The time column is extracted from the movie based on the rate frames are recorded (commonly about 30 frames per second or about .033 seconds between images), starting with t=0 at the start of the video. You need to define a new calculated time column that subtracts off the time value of your first digitized frame from the video time stamp so that you can plot using time values that start at t=0. This is essential to use our model equations, which assume the first data point corresponds to t=0.

3—Forces in equilibrium

Object

To gain experience in working with vectors graphically and algebraically, in the context of an object at rest and in equilibrium so that the net force acting on it is zero.

Background

If an object remains at rest Newton's 2nd law, $\Sigma \vec{F} = m\vec{a}$, tells us that the net force must be zero: $\Sigma \vec{F} = 0$. In this lab, given two specified forces you will find experimentally what third force is needed to keep a small ring stationary. This experimentally determined value will be compared with values found by graphical analysis, i.e., carefully drawing the vectors to scale, and by resolving the forces into components algebraically.

The forces on the ring are exerted by strings. Each string passes over a pulley and a mass, m, hangs at the end of the string. The tension the string exerts on the ring is then simply mg if the system is in equilibrium.

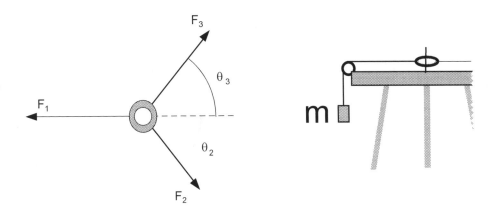

Figure 1: Forces acting on ring. Force table schematic.

Vectors and vector notation. A vector has both magnitude and direction. We denote a vector quantity symbolically by an arrow over it: e.g.,

$\vec{F}_2$. We can fully describe this vector with 2 (when restricted to a plane) or 3 (in three dimensions) numbers. For the 2-D case of this lab, we need two numbers, and these two numbers are commonly presented in two ways:
(1) magnitude (telling us the "length" of the arrow) and angle (telling us the direction it points): $\vec{F}_2 = (F_2, \theta_2)$;
(2) by the vector's components along x and y axes: $\vec{F}_2 = (F_{2x}, F_{2y})$. In component form sometimes the vector is written with unit vectors: $\vec{F}_2 = F_{2x}\hat{i} + F_{2y}\hat{j}$.

The magnitude F_2, *with no arrow*, is sometimes more formally written as $|\vec{F}_2|$. In either notation, the magnitude $|\vec{F}_2| = F_2 = \sqrt{F_{2x}^2 + F_{2y}^2}$ is always a positive quantity. In contrast, the components of a vector, F_{2x} and F_{2y}, may individually be positive or negative, depending on the direction of $\vec{F}_2$.

When referring to the component of a vector such as F_{2x} notice that there should be no arrow. Components of vectors are, by themselves, scalar quantities.

Unit vectors (denoted by the hat they wear) have a length or magnitude of 1 and are intended to convey direction only. Think of $\hat{i}$ as a shorthand way of writing "in the x direction" and $\hat{j}$ as meaning "in the y direction."

Procedure

Three forces. A card with your force table lists the magnitudes (in g) and directions of two forces $\vec{F}_1$ and $\vec{F}_2$. Your task is to find a third force $\vec{F}_3$ that keeps the ring on the table stationary and centered.

A note on units in this lab: Since each force involved is directly proportional to the mass hanging on the string, $F = mg$, you can for convenience in this lab record forces in grams. This is not correct practice in general, but the conversion to correct units (newtons) is in every case here accomplished by multiplying the mass (in kg) by $g = 9.80$ m/s^2. We will therefore omit this common conversion factor and—*for this lab only*—casually "measure" forces in grams.

Experimental determination of $\vec{F}_3$. Place the pin or nail through the ring into the hole at the center of the force table. This will keep the ring at the center of the table as you are setting up the forces on the ring for the equilibrium condition. Set one pulley at 180° for $\vec{F}_1$ and another at the angle given for $\vec{F}_2$. Put the strings over the pulleys and hang the specified masses on the ends of the strings.

To find the direction of the third force $\vec{F}_3$ required to keep the ring in equilibrium at the center of the table, grip the third string in your hand and by trial and error find the direction of pull (keeping the string parallel to the table surface) that will keep the ring at the center of the table. Place a third pulley at the required angle. To determine the magnitude of $\vec{F}_3$ place the third string over the pulley and by trial and error hang masses on the string to bring the ring to equilibrium at the center of the table.

When you are near equilibrium remove the pin from the ring. Friction in the pulleys will allow the ring to be in equilibrium over a considerable range of positions. To minimize the effects of friction test for the equilibrium position by lifting the ring about 2 cm up and releasing it. The ring should return to the *center* of the table. Test several times and find an average equilibrium position. Also be sure the three forces intersect at the center of the table: use a ruler to check carefully that each string's direction, when extended, will pass through the hole at the center of the table. You may slide the string on the ring to make this adjustment. Make final adjustments to the pulley position and mass if required. Record the magnitude and angle for $\vec{F}_3$. Make a quick sketch (not to scale) as in Fig. 1 of the forces showing the numerical values of F_1, F_2, F_3, θ_2 , and θ_3.

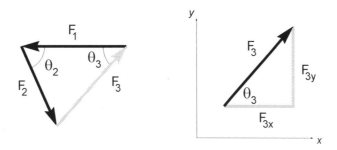

Figure 2: Forces summing to zero complete a triangle. Our choice of coordinate system for resolving $\vec{F}_3$ into components.

Graphical determination of $\vec{F}_3$. Three forces in equilibrium form a closed triangle when drawn head to tail in succession, the graphical method of finding $\vec{F}_1 + \vec{F}_2 + \vec{F}_3$. Using a clear sheet of unlined paper make a careful scale drawing as large as possible so that it still fits on a standard page. State explicitly what scale you are using, e.g. "1 cm = 500 g" (which is *not* a good choice) on your diagram. Start with a point near the upper right hand corner of the page and add $\vec{F}_1 + \vec{F}_2$ (fig. 2). $\vec{F}_3$ is found by closing the triangle by drawing a vector from the head of $\vec{F}_2$ back to the tail of $\vec{F}_1$. Measure the length of this vector. Use your scale factor to determine the magnitude of $\vec{F}_3$. Measure the angle θ_3. Record the graphical values of F_3 and θ_3.

Calculation of $\vec{F}_3$ by components. Resolve the three vectors into components. Let F_{3x} and F_{3y} be the x and y components of $\vec{F}_3$ Notice that $\vec{F}_1$ has only an x component, $F_{1x} = -F_1$ with the choice of axes in fig 2. Summing the x and y components and setting each equal sum to zero, we get

$$\Sigma F_x = F_{3x} + F_{2x} - F_1 = 0$$

and

$$\Sigma F_y = F_{3y} + F_{2y} = 0.$$

Solve these equations for F_{3x} and F_{3y} in terms of the magnitudes F_1, F_2, and θ_2. Using the values for F_1, F_2, and θ_2 given on your card calculate F_{3x} and F_{3y}.

Find the magnitude F_3 from $F_3 = \sqrt{F_{3x}^2 + F_{3y}^2}$ and θ_3 from $\tan\theta_3 = \frac{F_{3y}}{F_{3x}}$. Summarize your results by listing the experimental, graphical, and calculated values of F_3 and θ_3 in a table. In the table also list the percentage difference between the experimental and calculated values of F_3, and between the graphical and calculated values of F_3. Omit percent differences between the values of θ_3.

Four forces. Your card also lists a group of four forces. $\vec{F}_1$ now points along the $+x$ axis $(0°)$. The direction of $\vec{F}_2$ relative to $\vec{F}_1$, is also given, and illustrated as θ_{12} in Fig.$\tilde{3}$. Only the magnitudes for $\vec{F}_3$ and $\vec{F}_4$ are given, not their directions.

Find experimentally directions for $\vec{F}_3$ and $\vec{F}_4$ that produce equilibrium for the ring. Adjust the directions of $\vec{F}_3$ and $\vec{F}_4$ until equilibrium is achieved. Test for equilibrium as before, making sure the lines of force for all the strings pass through the center of the table. In a simple sketch of the arrangement list the four magnitudes and the three angles, as defined in Fig.$\tilde{3}$.

Graphically add $\vec{F}_1$ and $\vec{F}_2$ as before on a new sheet of plain paper. Choose a starting point and scale so that your diagram will be as large as possible and still fit on the page. The graphical determination of the directions for $\vec{F}_3$ and $\vec{F}_4$ requires some thought. While the magnitudes are known, the directions (ignoring the experimental determination just made) might be almost anything. The trick is to think about all the possibilities for adding $\vec{F}_3$ to the sum $\vec{F}_1 + \vec{F}_2$ that you've drawn already. From the head of $\vec{F}_2$ use a compass to draw a circle with radius equal to the magnitude of $\vec{F}_3$. The points on the circle represent all the possible outcomes of adding a vector with magnitude $|\vec{F}_3|$ to $(\vec{F}_1 + \vec{F}_2)$, but with any possible direction.

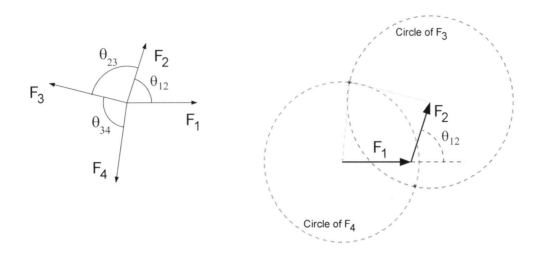

Figure 3: Example of forces, angles, and graphical construction for 4 forces.

We also know that the head of $\vec{F}_4$ must return to the tail of $\vec{F}_1$ to form a closed polygon representing the equilibrium case: $\Sigma \vec{F} = 0$. Working *backwards* from the starting point, the tail of $\vec{F}_4$ must lie on a circle of radius $|\vec{F}_4|$ from the tail of $\vec{F}_1$. Draw this circle with the compass. The intersection(s) of the two circles provides a determination of the required directions for $\vec{F}_3$ and $\vec{F}_4$. Draw in the appropriate choice for these last two vectors to close the polygon consistent with your experimental determination. Measure with a protractor the angles θ_{23} and θ_{34} on your diagram. Indicate which ones these are in your graphical construction but *be careful*. In this graphical construction these angles are exterior angles to the closed polygon; notice where θ_{12} appears in this construction. List the angles from your graphical construction in a table with the experimentally determined values for easy comparison.

Post-lab questions

The two circles drawn for the four-force situation generally have two points of intersection. You've probably naturally chosen one of them for the analysis just above. What physical situation does the second point of intersection correspond to on your force table?

Imagine three vectors satisfy $\vec{F}_1 + \vec{F}_2 + \vec{F}_3 = 0$. Now what can we reliably say about their magnitudes? Discuss which of the following two statements can ever be true for three such vectors.

$$|\vec{F}_3| > |\vec{F}_1| + |\vec{F}_2| \ ?$$

$$|\vec{F_3}| < |\vec{F_1}| + |\vec{F_2}| \ ?$$

Is it possible for three such vectors to satisfy

$$|\vec{F_3}| = |\vec{F_1}| + |\vec{F_2}| \ ?$$

Sketch an example of three such vectors, if so.

Before you leave: Clean up your lab table before you leave. Return all weights to their trays. Leave the tools for this experiment neatly arranged for the next section. Don't leave behind calculators, water bottles, random scraps of paper, etc.

4—Testing Newton's second law

Object

To investigate the relationship between force and acceleration represented by Newton's second law, $m\vec{a} = \vec{F}_{net}$.

Background

The famous relation between accelerations and their causes—forces—presented in Newton's second law can be tested experimentally. In this experiment, a large mass M (a glider on an air-track) is pulled by a small falling mass m. The air-track provides a very low friction surface for the glider to move on. The glider's motion can be measured with an ultrasonic motion detector. From graphs of $v(t)$ you can determine if the acceleration is constant and, if so, find its value.

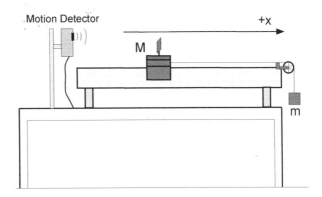

Application of Newton's second law to this system is straightforward. Draw separate free-body diagrams for the glider (M) and the falling mass (m), and apply Newton's second law to each.

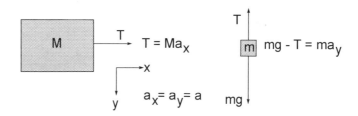

The diagram shown for M omits forces in the vertical direction: the weight of the glider and the balancing lift due to the air-track. The coordinate system is drawn so that the horizontal acceleration of the glider and the downward acceleration of m are both positive. Both must have the same value—otherwise the string must change length! Only the string pulls horizontally on the glider, with a tension T. The falling mass is subject to two forces, the tension of the string and its weight (the gravitational pull exerted by Earth). Newton's second law, $\Sigma \vec{F} = m\vec{a}$, is applied to each object. The two equations that follow from the free-body diagram can be used to eliminate the tension, T, from the equations, since we will not measure that. The equation for the falling mass can be re-written without T:

$$mg - T = ma$$
$$mg - Ma = ma$$
$$mg = (M + m)a \qquad (1)$$

This is solved for the acceleration, a, of the glider (and of the falling mass):

$$a = g\left(\frac{m}{M + m}\right) \qquad (2)$$

One way to interpret this result to look at Eq. 1. This can be seen as stating that the net *external* force (mg) on the "system" (glider and falling mass) equals the mass of this combined system ($M + m$) times its acceleration.

   The results of the measurements can be graphed to see if Eq. 2 is supported by experiment. If we regard the quantity $X = m/(M + m)$ as the independent variable (the quantity that we deliberately change when carrying out the experiment), Eq. 2 is simply the equation of a straight line passing through the origin: $a = gX$ and a graph of the data – $(\frac{m}{M+m}, a)$ values – should fall very nearly along a straight line if our original assumption — $F = ma$ — is correct.

   According to Eq. 2, for a given combination of m and M, the acceleration is constant. Then the glider's motion should be described by formulas familiar from one-dimensional kinematics. A graph of $v(t)$ should be linear with a slope corresponding to the acceleration: $v(t) = at + v_o$. You will use the motion detector to determine if $v(t)$ is linear and, if so, find the acceleration from the slope.

Procedure

Set-up. The air-track and especially *the gliders* should be handled very carefully. The gliders can be bent easily and will no longer fit the air-track properly.

- Attach a string to one glider and adjust its length so that when the glider reaches the spring bumper at the end of the air-track, the string with a washer hanging on it will just reach the floor. When the glider is pulled back up the track, it should reach about 100-110 cm from the bumper. The string must be parallel to the track and pass smoothly over the pulley without rubbing on any parts of the track.

- Level the air-track by turning the two adjustable feet toward one end of the track simultaneously until the glider moves no more than 1 cm in 4 or 5 seconds when released from rest. The air-track and blower should be warmed up before leveling and should remain on until measurements are complete. The change in temperature in the air-track from turning the blower on and off causes thermal expansion of the track and negates your efforts at making it level.

- Set up the motion detector to track the glider's motion. Recall that the motion detector does not work correctly for objects closer than about 45 cm. Place the motion detector about 50 cm behind the back end of the glider at your release release point.

- The file "newton2nd.cmbl" is configured to collect data for this experiment. Open it and make sure you have aligned the motion detector properly to track the glider over the full range it will move when the string is attached. This experiment file is configured to store data only *after* the glider has begun moving *and* traveled past a particular distance from the detector (55 cm). This makes successive runs easier to compare when you use a consistent release point. You will hear the detector operating, but may not see data being collected immediately. Don't expect your resulting $v(t)$ curves to start at $v = 0$.

Single glider with 1, 2, and 3 washers. Your task is to determine if the acceleration is constant while the glider is being pulled by one or more falling washers hung on the string. Begin with a single washer. Find its mass and the mass of the glider.

- Record the motion of the glider with the motion detector as it is pulled along the air-track by the falling washer. Start the data collection but wait 1-2 seconds before releasing the glider. Keep track of your release point so you can reproduce your trial.

- Does the $v(t)$ graph support constant acceleration prior the glider reaching the bumper or the washer touching the floor? What aspect of the graph leads you to that conclusion? Is the $x(t)$ graph generally what you expect?

- If the graph shows constant acceleration, use the regression tool from the menu bar to analyze the data. Select the data in the graph that you believe is consistent with constant acceleration and use the linear fit tool button on the menu bar. This automatically computes the best straight-line fit to the selected data using the least squares method (see the appendix on this topic). The corresponding best slope and intercept in the correspondence between $y = mx + b \iff v = at + v_o$ are displayed. Record the value of the acceleration reported by *Logger Pro* in your notebook, along with m and M. Be sure to include correct units.

- Store the data from this run (CTRL-L or 'Data – Store Data Set – Latest' from the menu bar) for comparison with subsequent trials.

- Make at least two more trials with this configuration (1 glider, 1 washer). Use the same release point. Are your runs consistent? If not, is there reason to discard a trial? Discard a trial only if you have good reason, for example if the motion detector failed to follow the glider accurately, or the washer fell off before the run was complete. Store the data and carry out the linear fit analysis to get and record the acceleration for each successful trial.

- Repeat the entire process (three successful trials and linear fits) with two washers pulling the glider. Weigh the pair of washers so you know the total falling mass, m. Measuring each individually and adding combines errors of separate measurements and is less precise. Keep the data from the one-washer trials in Logger Pro so you can compare the two cases on the same graph. Use the same release point.

- Repeat again with three washers pulling the single glider.

- Record all accelerations from fits in your notebook and close any boxes displaying fit results on the graph. Print out the single graph containing all 9 $v(t)$ curves, presumably in three distinct clusters if you have been consistent about your release point, from all three combinations for inclusion in your notebook. Save the data so you can re-examine it later if necessary. You can then continue on in *Logger Pro* with a clean slate by choosing 'Data - Clear All Data' from the menu bar.

Two gliders. Repeat the measurement and analysis cycle using two gliders coupled together with Velcro pulled by 1, 3, and 5 washers. Be sure to record the accelerations, total glider mass and the washer masses for each case. Again include the single graph containing all nine $v(t)$ curves for these trials with two gliders in your notebook. Save the *Logger Pro* data for later review!

Analysis

The measurements and analysis made to this point should demonstrate that the acceleration during a trial is constant, but does the large collection of accelerations gathered support Newton's second law? Test the validity of Newton's second law by graphing your accumulated acceleration data in a way that tests Eq. 2. Assuming your data supports the claim of constant acceleration motion, Eq. 2 predicts a particular relationship between the measured acceleration and the masses M and m.

Construct a graph of $(X, Y) = (\frac{m}{M+m}, a)$ values. Plot all 18 trials. You can calculate $\frac{m}{M+m}$ with a calculator and enter those values and the accelerations directly into a table in a basic *Logger Pro* file - "manual_graph.cmbl" provides an easy starting point for manually entering and graphing data. Customize the column names and units as needed.

Do the data plotted in this way form a straight line? If so, then Newton's second law is supported by the experiment. Find the "best-fit" straight line through the data using the fitting tool in *Logger Pro*.

According to Eq. 2, what should the slope of this line be? What should the 'Y'-intercept be? Make this comparison clear by constructing a line on your graph that represents Eq. 2, with the predicted slope. Include a printed copy of this graph in your notebook.

Additional analysis:

①　(a) Do you expect the tension in the string as the system accelerates to be less than, equal to, or greater than the weight of the falling washer(s) mg? Explain why you think so.

(b) Determine from your data the tension in the string from the glider mass and its acceleration for the case of one glider and three washers. Compare this to the weight of the falling washers. Are the values consistent with your expectations?

The slope of your line fitted to all 18 accelerations vs. $m/(M + m)$ may differ from g by more than can be attributed to the uncertainty in a values, and the line may not pass through the origin. These differences may reflect *systematic errors* in the experiment. Our simple model has neglected two small but possibly significant effects: the presence of drag or frictional forces and the rotational motion of the pulley.

(2) The drag forces reduce the net external force causing the acceleration. This can be accounted for semi-quantitatively by modifying Eq. 1, replacing mg with $mg - F_d$. The drag or frictional force (F_d) can be written more conveniently as $F_d = m_d g$, where m_d would be the hanging mass needed to pull the glider at constant velocity ($a = 0$) along the track, just balancing the drag forces. Eq. 2 becomes $a = g\left(\frac{m - m_d}{M + m}\right)$

(a) The X-intercept of the graph just constructed (where $a = 0$) can provide an estimate of m_d and the drag force for each of the two glider masses: $X_{int} = m_d/(M + m)$ so that $m_d = (M + m)X_{int} \approx MX_{int}$ (since $m \ll M$). With two distinct Ms there are two different values of m_d you can estimate. Figure a single value value for m_d using an average M for the two cases.)

(b) Describe how you would experimentally determine m_d, with the equipment of this experiment plus a collection of paper clips, which are quite a bit lighter than a single washer. Be specific in your experimental technique and how you would decide on the correct m_d.

(3) The rotational motion of the pulley cannot be properly handled with techniques learned so far. A simple-minded correction would be to account for the mass of the pulley as part of the accelerating system, replacing the denominator in Eq. 2 with $(M + m + m_p)$ where m_p is the effective mass that the pulley contributes to the accelerating system. Eq. 2 then becomes

$$a = g\frac{m}{M + m + m_p} \tag{3}$$

(Including the drag/friction effects at the same time would make the numerator $m - m_d$ instead of just m.) Treat $\frac{m}{M + m + m_p}$ as your new X. Try $m_p = 5\,\mathrm{g}$ and recompute these values and re-plot your graph. Check that the data still fall close along a straight line and evaluate the new slope. Is it any closer to g?

Can you find a value of m_p that brings your slope closest to $9.80\,\mathrm{m/s^2}$? Is that value of m_p plausible?

5—Estimating experimental uncertainties

Object

To estimate the uncertainty in the measurement of the density of a solid. The treatment of experimental uncertainties is a central technique to be gained from this lab. You will also learn to use micrometers and calipers with a vernier scale for making more precise measurements. Read the appendix on error analysis before coming to lab.

Background

Density, ρ (the Greek letter "rho"), is defined as the amount of mass contained in a unit volume. A simple way to determine the average density of a solid is to use a simple geometrical shape so its volume can be easily found. Find the mass of the sample using a balance, and then calculate the ratio of mass to volume, $\frac{M}{V}$, to obtain the density. In S.I. units the unit volume is one cubic meter, (m^3) and density is expressed in kg/m^3. More convenient units for density use smaller unit volumes. For example, densities are often given in g/cm^3. Since it's possible to lift a few cubic centimeters of a material in your hand and crudely compare the mass to other samples, these units offer a more human scale for thinking about and comparing densities. It's difficult to imagine lifting (by yourself) a cubic meter of iron, but a hefting a few cubic centimeters in palm of your hand is in the realm of possibility.

In this lab the initial sample is a cylinder of mass M, diameter d, and length L. Its volume, V, is its length times its circular cross-sectional area: $L \cdot \pi(\frac{d}{2})^2$. Its density can be found from

$$\rho = \frac{M}{V} = \frac{4M}{\pi d^2 L} \text{ave.}$$

If the mass is measured in grams, and L and d in centimeters, the resulting units are g/cm^3. Typical densities of ordinary solids are roughly 1-20 g/cm^3.

No measurement can be made with perfect precision or accuracy. There is always some uncertainty or "error" determined partly by the skill and patience of the experimenter, and partly by the accuracy of the apparatus. Although these effects are often called errors and their treatment called error analysis, keep in mind that this doesn't mean that something was (necessarily) done wrong; it's really the limitations and uncertainties in the measurement process we seek to estimate. The careful experimenter will estimate the uncertainties or probable errors in his or her measurements and determine the effect of these uncertainties on the final result. Ultimately, this final uncertainty should indicate the what range of results would be obtained by others if they repeated the experiment with the same kinds of measuring tools.

Let ΔM, Δd, and ΔL be the estimates of uncertainty in the measured values of the mass, diameter, and length of the sample respectively. This means the experimenter is reasonably confident the true mass lies between $M - \Delta M$ and $M + \Delta M$, the actual diameter falls between $d - \Delta d$ and $d + \Delta d$, and the actual length lies between $L - \Delta L$ and $L + \Delta L$. (Note that Δ here does not indicate a *change* in a quantity.)

In expressing the uncertainties in the individual quantities, the fractional error (uncertainty) is a useful quantity. The fractional error is the uncertainty divided by the mean value of the measurements (considered the best estimate of the true value). We'll denote the fractional error in a quantity x as $e_x = \frac{\Delta x}{\bar{x}}$. Sometimes it's useful to think of this as a percentage of the measurement:

$$\%\text{uncertainty in x} = \frac{\Delta x}{\bar{x}} \times 100\%$$

So if $x = 2.00 \pm .04$ cm, the fractional uncertainty is 0.02 or 2%.

Since the density ρ is calculated from these values, the density itself must have an uncertainty $\Delta\rho$ that depends on the uncertainties in M, d, and L. For quantities that are *multiplied* or *divided*, the fractional errors are especially useful as the fractional undertainty in the result will be estimated from the *sum* of the fractional uncertainties of the original measurements. (See the appendix on error analysis.) The resulting fractional uncertainty in the density is given approximately by

$$\frac{\Delta\rho}{\rho} = \frac{\Delta M}{M} + \frac{\Delta V}{V}$$

and

$$\frac{\Delta V}{V} = 2\frac{\Delta d}{d} + \frac{\Delta L}{L},$$

so

$$\Delta\rho = \rho\left(\frac{\Delta M}{M} + 2\frac{\Delta d}{d} + \frac{\Delta L}{L}\right).$$

Procedure

Mass measurement. Note whether the cylinder you select appears to be steel, brass, copper, or aluminum. Make at least one determination of the mass, M, of your cylinder with each balance in the lab room. Each member of the lab team must take a turn in making some of these measurements. Zero the balance before making the measurement and record each mass value to the resolution that the balance provides.

#'s to micrometer
±0.03

You should collect your measurements in one or more neatly drawn tables.

Trial	Mass (g)	Deviation	Length (cm)	Deviation	Diameter (mm)	Deviation
1	...	...	...	...	...	...
⋮	⋮	⋮	⋮	⋮	⋮	⋮
n	...	...	...	...	...	...
Avgs	$\bar{M} =$	$\sigma_M =$	$\bar{L} =$	$\sigma_L =$	$\bar{d} =$	$\sigma_d =$

Calculate the average or mean, $\bar{M}$, of all the mass measurements (M_i, $i = 1 \ldots n$). This average is the best estimate of the mass to use for calculating the density. Calculate also the deviation ($|M_i - \bar{M}|$) of each measurement from the average mass. Calculate the standard deviation for this set of measurements:

$$\sigma_M = \sqrt{\frac{\sum (M_i - \bar{M})^2}{n - 1}}$$

using a calculator, preferably with statistics functions. This standard deviation will be a reasonable estimate of the uncertainty ΔM provided there is some variation in the measurements. If all measurements are the same, the standard deviation would be zero. This does not mean there is no uncertainty in the measured mass; instead, the uncertainty must then be estimated from the resolution of the balance or additional information from the manufacturer about its accuracy. In the absence of any additional information we will assume the minimum uncertainty to be one-half the least significant digit of the balance reading.

> *We will adopt the following rule:* If the statistical uncertainty, *e.g.*, σ, of a set of measurements is more than half the smallest division of the measuring instrument, use the statistical uncertainty as the estimate for the uncertainty in the measured quantity; otherwise take the uncertainty in the measurment to be one-half the smallest division of the measuring instrument.

Choose for your value of uncertainty ΔM the **larger** of σ_M and the instrumental limit. Record your choice and your justification in your notebook.

Report the fractional uncertainty in the mass, too.

Length measurement. Using vernier calipers, make at least $n = 6$ determinations of the length of the cylinder. Determine the resolution of the vernier scale—the fraction of a millimeter represented by the smallest sub-division of the vernier scale. (A vernier scale with a total of 50 subdivisions provides resolution to $1/50$ mm or $.02$ mm. Record each reading to this precision. Be sure to make each measurement independent, removing the cylinder from the calipers in between measurements. Each member of the lab team must make some of these measurements and understand how to read the calipers. Calculate the average of the readings, $\bar{L}$ and the standard deviation, σ_L. Compare σ_L to the resolution limits of the calipers and choose the larger as your estimate of the uncertainty, ΔL. Record your choice and your justification in your notebook, along with the corresponding fractional uncertainty.

Diameter measurement. Using micrometer calipers (usually called simply a micrometer), make at least $n = 12$ determinations of the diameter of the cylinder. Pick distinct locations again: the cylinder may be non-uniform in diameter along its length or slightly out of round. Note that the micrometer is calibrated in millimeters. To avoid confusion, record your readings in mm and convert afterwards to cm. Notice also that the drum of the micrometer has 50 divisions on its circumference. The drum advances 1 mm for every *two* complete turns, so 100 divisions on the drum represent 1 mm and 1 division is 0.01 mm. Also notice that you need to add 0.50 mm to the visible mm line whenever the drum is in the second half of a millimeter division.

Bring the jaws of the micrometer together *gently* against the cylinder! Do NOT overtighten the micrometer. Record each reading to the nearest 0.01 mm.

After making all the measurements, record the zero reading of the micrometer with the jaws brought together so there is zero distance between them.

Do NOT over-tighten the jaws. Do not force the reading to zero. This reading may be positive (e.g., the micrometer reads 0.03 mm when the jaws are brought gently together, or must be taken as negative if the micrometer jaws close after passing through zero, in which case you'll need to count th number of divisions past zero you've gone e.g., a reading on the barrel of 47 corresponds to a correction of -0.03 mm In either case the corrected diameter reading is given by

```
(true diameter) = (micrometer reading) - (zero reading).
```

Remember that this means if the zero reading is negative, the true diameter is larger than the micrometer reading. Failure to take into account this zero position would constitute a systematic error in your measurements. Since our diameter measurement involves subtraction two measurements, the minimum uncertainty in the diameter is 0.01 mm, not 0.005 mm. The minimum instrumental uncertainty in both the reading of the diameter and in the zero reading of 0.005 mm, *add* to provide a total uncertainty in the true diameter of at least 0.01 mm.

Find the average of the twelve diameter measurements, $\bar{d}$, and the resulting standard deviation, σ_d for the measurements. Again select the appropriate estimate of the uncertainty in the diameter, Δd, based on a comparison of the statistical uncertainty σ_d and the minimum uncertainty of 0.01 mm imposed by the micrometer's resolution. Record your choice and your justification in your notebook. Report the fractional error, too.

Density and propagation of errors calculations. State clearly what average values you found for M, d, and L for your cylinder. Use these to find your best estimate for the density of your cylinder.

Summarize clearly what values you have decided to use for the uncertainties ΔM, Δd, and ΔL. Compute and list the *fractional* error or uncertainty in each of these measured quantities, $\Delta M / \bar{M}$, $\Delta d / \bar{d}$, and $\Delta L / \bar{L}$. Use these and the value for the density you report to determine the uncertainty in your value of ρ using the prescription in the Background section for $\Delta \rho$. Note that this is only an estimate and you should round this value off to 1 or 2 significant figures.

Given this uncertainty, how many significant figures are justified in your calculated value of the density itself? It is common (in the absence of more sophisticated statistical analysis) to round off experimental results to the point where the result has only one digit that is uncertain. For example, if the density was found to be $8.551 \pm .032$ g/cm^3, it would be rounded and reported as $8.55 \pm .03$ g/cm^3, because we are fairly confident of the first two digits but less so of the third. Any more digits are so uncertain they should not be reported in your final results. What is the fractional uncertainty in your density measurement?

Second cylinder. Repeat the measurements and calculations for a second cylinder of a different material.

Compare your results with the following values of density from standard tables and identify the material of your cylinder. The uncertainties in these reported values reflect the range of densities encountered due to variations in composition, methods of treatment, and perhaps other factors, not intrinsic uncertainties from a particular measurement process. Your average value may fall within the range indicated, but your uncertainty may be quite different from the ranges listed in the table.

Aluminum	$2.70 \pm .03$ g/cm^3
Brass	$8.44 \pm .06$ g/cm^3
Copper	$8.87 \pm .06$ g/cm^3
Steel	$7.83 \pm .08$ g/cm^3

Marble. Determine the density and its uncertainty (and fractional uncertainty) for a marble provided by your lab instructor, using a dozen independent measurements of its diameter. (The volume of a sphere is $\frac{4\pi}{3}r^3$.)

Before you leave: Make sure your lab table is ready for the next group. Leave equipment neatly organized for the next lab. In particular, be sure the micrometer is stored in its case with its jaws slightly apart, *not* closed together. Please don't leave random scraps of paper, calculators, water-bottles, etc., behind.

Post-lab questions. Answer with complete sentences.
(1) Which quantity (mass, length, or diameter) had the largest fractional or percentage uncertainty for each of your cylinders.
(2) Which quantity was most influential (contributed the most) in determining the uncertainty in the density of your cylinders?
(3) Apply the "Crank-three-times" rule in the appendix to your first cylinder to find its density. How does the range of uncertainty obtained from this compare to your previous "error bars?"
(4) Which quantity (mass, diameter) is most influential in determining the uncertainty in the density of your marble?

Statistical calculations and Logger Pro. It is a useful exercise to learn how to use the statistical calculation features your calculator probably has built into it. It may take some experimentation (or a few minutes with the user's manual) to discover what options to select to get the correct estimate of the standard deviation for a small number of samples:

$$\sqrt{\frac{\sum (x_i - \bar{x})^2}{n-1}} \text{ instead of } \sqrt{\frac{\sum (x_i - \bar{x})^2}{n}}$$

which is correct version for a very large (or infinite) number of measurements. Some calculators may denote the first, appropriate for our use, by s_x, while the calculation that divides by n might be denoted by σ_x. That 2nd form is valid only when the number of measurements becomes large (so that dividing by n or $n-1$ doesn't make much of a difference).

Logger Pro can calculate statistics of a set of measurements, using the correct (i.e. dividing by $n - 1$) formula. You manually enter your measurements into columns and then make a graph with the quantity you want statistics on as the y-axis. (You might make a dummy x-axis column that just runs from 1 to the number of meaasurements.) Selecting the data in the graph with the mouse and using the statistics button will give a report on the mean and the proper estimate of the standard deviation.

6—Hooke's law and simple harmonic motion

Object

To measure the spring constants of springs, measure the period of oscillation of a mass on a spring, and determine the parameters of simple harmonic motion that describe the oscillations of the mass.

Background

Hooke's law describes the stretching of a spring. Let s be the increase in length of a spring from its unstretched state when a force F pulls on the spring. According to Hooke's law, the magnitude of the applied force and the amount of stretch are proportional: $F = ks$. The proportionality factor k is called the spring constant. You will measure k using two methods.

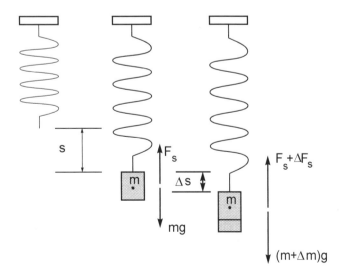

A mass, m, hangs vertically from the end of the spring. The mass is subject to two forces: the upward force exerted by the spring whose magnitude is proportional to the amount s the spring has stretched ($F_s = ks$), and the gravitational pull exerted by the earth, mg downward. When the mass is at rest (in equilibrium) the forces must be equal in magnitude: $F_s = mg$. If the mass is increased, the spring stretches more:

$$F_{s1} = ks = mg \qquad\qquad F_{s2} = k(s + \Delta s) = (m + \Delta m)g$$

Subtracting these two and solving for k gives

$$F_{s2} - F_{s1} = k(\Delta s) = (\Delta m)g,$$

$$k = \frac{\Delta F}{\Delta s}\left(= \frac{(\Delta m)g}{\Delta s}\right). \tag{1}$$

Equal increments in the mass should lead to equal increments in the length of the spring. When forces are measured in newtons and lengths in meters, the units of k should be N/m. According to Hooke's law, a graph of F_s as a function of s should be linear with slope k.

When the mass is displaced from its equilibrium position—either by pulling it down or lifting it up—and released, it oscillates. The force exerted by the spring and the weight are no longer equal. The net force acting on the mass is the difference between the spring force and the gravitational force and is simply $-ky$, where y is the displacement from equilibrium. The negative sign reflects the fact that the net force is upwards when the mass is below equilibrium—stretched more—and downwards when above equilibrium—stretched less. The force and displacement are in opposite directions. Newton's 2nd law, written as $a = F/m$, with $a = \frac{d^2y}{dt^2}$) then gives

$$\frac{d^2y}{dt^2} = -\frac{k}{m}y.$$

The resulting oscillatory motion $y(t)$ of the mass is called simple harmonic motion and is sinusoidal in time:

$$y(t) = A\sin(\omega t + \phi), \tag{2}$$

A is the amplitude, the largest distance from equilibrium that the mass achieves. $\omega = \sqrt{\frac{k}{m}}$ is called the angular frequency. Experimentally it is easier to measure the period of oscillation, T, which is the time required for one complete cycle of the mass's motion:

$$T = \frac{2\pi}{\omega} = 2\pi\sqrt{\frac{m}{k}}. \tag{3}$$

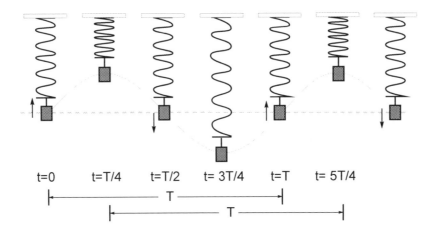

The period is evidently independent of the amplitude of the motion, depending only on the mass and the spring constant. If the mass of the spring is not negligible compared to the added mass at the end, a more complete treatment shows the mass of the spring influences the period with

$$T = 2\pi\sqrt{\frac{m + \frac{m_s}{3}}{k}}.$$

And this means that a more precise expression for ω becomes $\sqrt{k/(m + (m_s/3))}$. The phase ϕ determines the position and velocity at $t = 0$, shifting the sine wave curve left or right.

Procedure

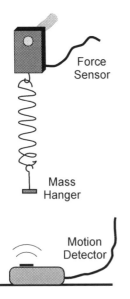

The measurements will be made using a force sensor to measure the pull exerted by the spring and a motion detector to measure the stretching or oscillatory motion. The motion detector must be connected to a DIG/SONIC input and the force sensor to an analog input channel, e.g., CH 1, of the LabPro interface. Set the force sensor for the ± 10 N range.

Spring constant: static measurement. The starting point for measurements is contained in the *Logger Pro* experiment file "hooke_static.cmbl." A graph of the force as a function of distance is produced. In this file, a positive force reported by the force sensor corresponds to the spring exerting an upward force on the mass. The distance reported by the motion detector is based on a coordinate system that has an origin set arbitrarily about 1.2 m above the motion detector. The coordinate system increases downwards. So $y = 0$ is a point 1.2 m above the motion detector, and the motion detector is at $y = +1.2$ m. This means the distance reported by the motion detector increases as the stretch of the spring increases. In a sense, the sign conventions used to report the spring's force and stretch are opposite: a positive force is upward, while a positive stretching is downward. This merely gives graphs made below positive slope for convenience.

Force calibration. With the force sensor mounted on the supporting rod, hang 400 g directly from the sensor—no spring. (Count the mass hanger.) Be sure the sensor is oriented so the mass pulls vertically down on the sensor's mounting hook. A (tiny) live force readout is in the status bar at the top of the *Logger Pro* window. The reading should be very nearly 3.92 N. (You can insert a larger, easier to read display: 'Insert - Meter - Digital.' Right-click on the meter to bring up options to select the latest force sensor reading.)

If the difference is more than .05 N, check the calibration of the force sensor: Click "Experiment - Calibrate," select the force sensor, and choose the "Calibrate" tabbed page. Click "Calibrate Now." You need to provide two known points to calibrate the sensor. Remove all weights from the sensor, and with nothing hanging from the sensor enter 0.0 in the "Reading 1" box, then click "Keep." Next hang a total of 400. g from the sensor and enter 3.92 N ($= mg$) as the second calibration value, "Keep." Exit the calibration process.

Small steel spring. Hang the spring from the sensor. Note the force and make a very rough estimate the mass of the spring. Check the mass with a balance. Place the motion detector and protective basket directly beneath the spring. Add the 50 g mass hanger to the spring. Adjust the height of the support rod so that the mass hanger's bottom is about 1.0 m above the motion detector. Make sure the motion detector is directly beneath the mass hanger and they are positioned away from the edge of the table to avoid spurious echoes from the table.

When you start data collection with the "Collect" button, *Logger Pro* will begin to sample the force and distance repeatedly, displaying each point briefly on the graph but not saving the data. Make sure the distance value is reasonable and not due to stray reflections of the ultrasound. Once the mass has reached equilibrium and motion has damped out, click the "Keep" button to store measurements in the data table. Store four measurements, collected over roughly 10 seconds to provide some averaging of the residual vibrations and noise. Don't press "Stop" yet.

Add 20 g to the hanger (70 g total). After the mass stabilizes continue data collection, taking four independent measurements at this mass. Repeat for total masses of 90, 110, 130, 150, and 200 g. (You'll need to add a larger diameter mass to reach the final value.) Press "Stop" when all masses are complete.

Use the linear fit tool to find the slope of the best-fit line through your data. <u>What does this slope represent? What are the units?</u> Store the data (Ctrl-L) for later printing.

<u>Large spring.</u> Repeat the process for the larger brass spring. You may need to re-adjust the height of the supporting rod so that the mass hanger again starts out around 1 m above the motion detector. Use total masses of 50, 100, 150, 200, 250, and 300 g. Find the spring constant for this spring.

Print out a graph showing the data and fits for both springs. Save the data for later review.

Spring constant: dynamic measurement. Open the *Logger Pro* experiment file "hooke_dynamic.cmbl." This file also measures force and distance, and at a high enough rate to measure the values as the mass is oscillating up and down. With the new file open, verify the calibration of the force sensor again.

With 300 g on the large spring, set it in motion by pulling the mass down 10-12 cm and releasing. Start data collection. Data is collected at 20 samples per second for 10 seconds over several complete cycles. When complete, fit the entire data set to a straight line via "Analyze – Linear Fit" (instead of using the regression button) and get a value for the spring constant. Store the data.

Mount the smaller spring back on the force sensor and adjust the supporting rod height again to make the mass hanger alone reach to about 1 m above the motion detector. With 100 g total hanging, set the system into oscillation and collect data. Find the spring constant by fitting the entire data set again when measurements are complete.

Print out a single graph comparing data from both springs for your notebook. Save the data for later review.

Period measurement. Use the small spring and a 50 g load. Open the experiment file "shm_1.cmbl." This file displays in separate panes of a graph window the position, velocity and acceleration as functions of time. In this file, both position of the mass and the force exerted by the spring on the mass are measured with upwards as positive.

While the mass is stationary at the equilibrium point, zero the motion detector and force sensor: select "Experiment - Zero - Zero all sensors" or use the "Zero" button. *Logger Pro* samples the current distance and force values, stores them, and automatically subtracts them from subsequent measurements. This makes the equilibrium position the origin of position measurements, and subsequent force readings will reflect the *net* force on the mass (which is zero at equilibrium).

Pull the mass down 3 to 5 cm and release. Collect data. Data is collected automatically for 10 s. From the graph measure the period of oscillation: Use the coordinate read-out tool from the button bar ("Examine") to determine the first time $y(t)$ reaches a maximum, t_0. Determine the time for the last maximum, t_N. How many complete cycles, N, does this interval contain? Find the period of one cycle: $T = (t_N - t_0)/N$. Store the data.

Measure the period for $m = 100$ g and 150 g. Re-zero the sensors for the new equilibrium each time. To plot only the data of the latest run, click on the vertical axis label of a graph and make sure only data for the latest run is checked. Store the data for each m before starting the next. Save all the data to a file for later review.

Summarize your results in a table in your notebook:

m	t_0	t_N	N	T
...	...	...	...	...

and in a graph: change the data displayed (click on the vertical axis label and check and uncheck choices as needed) in the three panes of the graph window to show only $y(t)$ for each run in a separate pane:
  • Top pane: $y(t)$ of latest run ($m = 150$ g)
  • Middle pane: $y(t)$ of run 2 ($m = 100$ g)
  • Lower pane: $y(t)$ of run 1 ($m = 50$ g).
Print out the graph window for your notebook.

Reset the graph panes to show y, v and a for the latest run only. Examine the resulting curves to answer the following questions.
  • Where is the mass when its velocity reaches a maximum? When v is minimum?
  • Where is the mass when the velocity is zero?
  • Where is the mass when its acceleration is most positive?
  • Where is the mass when the acceleration is zero?

Determining amplitude, angular frequency, and phase. The graph of $y(t)$ for the last run can be described by the oscillation of Eq. 2 with (perhaps) an additional offset:

$$y(t) = A \sin(\omega t + \phi) + y_o. \tag{4}$$

Logger Pro can superimpose a curve with similar mathematical form over the data. Click on the $y(t)$ graph pane to select its data. Pick a function to superimpose through the "Analyze - Curve Fit." Select the manual curve fit option, not automatic.

From the available choices of general equations select the sine equation:

$$A * \sin(B * t + C) + D.$$

The correspondence of parameters between our $y(t)$ and the *Logger Pro* function should be clear:

Parameter Name	Eq. 4 $A\sin(\omega t + \phi) + y_o$	Logger Pro $A * \sin(B * t + C) + D$
	t	t
Amplitude	A	A
Angular frequency	ω	B
Phase	ϕ	C
Y-Offset	y_o	D

You can estimate the amplitude from your $y(t)$ graph: the distance from the highest to lowest points on the curve should be $2A$. You can also estimate $y_o = D$ by looking at the overall average y. Since you have measured the period, estimate $\omega = 2\pi/T$. Enter these values into the parameter boxes in *Logger Pro*. Adjust last parameter, the phase ($\phi = C$), by hand. Increment the value and watch what happens to the fit curve. Once you have roughly the correct values for each parameter, you can fine tune them[1] and observe how each one affects the calculated curve relative to the collected data. Record the final values of your parameters.

Print out the graph page, including the fit to $y(t)$ and the $v(t)$ and $a(t)$ plots.

Questions.

1. (In lab) Measure the force-distance curve for a rubber band. Use the dynamic method and simply slowly stretch and relax the rubber band by hand. What can you say about the work done by the rubber band on stretching and contracting?

2. Using the spring constant, k, found from the static method, compare the measured periods to the values predicted from Eq. 3. What should a graph of T^2 vs. m look like? Does it? How significant a correction is the spring's mass?

---

[1]If the step size by which *Logger Pro* changes a parameter is too coarse and produces too large a change in the fitting curve, click on the tiny Δ to enter a smaller step size.

3. Differentiate Eq. 4 to find an equation for $v(t)$. Use the parameters found from fitting the oscillating $y(t)$ data to create a new calculated column using this equation. Compare the calculated $v(t)$ with the measured data.

4. Create new calculated columns of the kinetic, potential, and total energies using the data from the last run of the period measurements. Graph them. How well does this system exhibit conservation of energy? Are there any significant discrepancies in any of the curves?

7—Atwood's machine and energy

Object

To explore the conservation of mechanical energy by measuring kinetic, potential, and total mechanical energy in Atwood's machine.

Background

The kinetic energy of an object with mass m moving at a speed v is $K = \frac{1}{2}mv^2$. A small change dK in the kinetic energy of m as its speed changes by dv in a short time interval dt is related to the force accelerating the mass. For motion along a straight line (one dimension) starting from our definition of K we can write:

$$
\begin{aligned}
dK &= mv\,dv \\
&= mv\,a\,dt \ \text{ since } a = \frac{dv}{dt} \\
&= F\,v\,dt \ \text{ pulling together } ma = F \\
dK &= F\,dx \ \text{using } v = \frac{dx}{dt}.
\end{aligned}
$$

The total change in kinetic energy is found from the integral of the force along the displacement (from starting point i to ending point f) and is the work, W, done by the force on the mass:

$$
\begin{aligned}
K_f - K_i &= \int_i^f F\,dx \\
\Delta K &= W.
\end{aligned}
$$

This is the "Work—Kinetic Energy Theorem."

A force is described as *conservative* if it can be written in terms of the derivative of another function: $F_x = -\frac{dU}{dx}$. The function U is called the potential energy. From the definitions of work and potential energy, an infinitesimal displacement dx involves work $dW = F_x dx = -dU$. The total

61

work done by F_x as m moves from x_i to x_f is $W = -\Delta U$. For purely conservative forces this means that $\Delta K = W = -\Delta U$ or $\Delta K + \Delta U = 0$.

For the case of the gravitational force on a mass near Earth's surface, $F_y = -mg = -\frac{d(mgy)}{dy}$, where y is the vertical position of the mass measured relative to some convenient origin, increasing upward. Gravitational potential energy near Earth's surface can then be written as $U(y) = mgy$.

The total mechanical energy is defined as $E = K + U$. If all the forces acting within a system are conservative—i.e., can be associated with potential energy functions—the total energy remains constant as the parts of the system move: $\Delta K = -\Delta U \implies \Delta K + \Delta U = \Delta E = 0$. This is the principle of conservation of mechanical energy.

Many situations include non-conservative forces. A common example is friction; another example could be a external agent, like a person, pulling or pushing an object whose total energy is not included as part of the system studied. When forces are a mixture of conservative and non-conservative forces, the work done by the various forces can be divided into two pieces: a conservative contribution accounted for by potential energy, and non-conservative work W_{nc} so that the toal work $W_{total} = -\Delta U + W_{nc}$ The presence of non-conservative work will lead to a change in the total energy equal to the work done by non-conservative forces:

$$\Delta E = W_{nc}.$$

Procedure

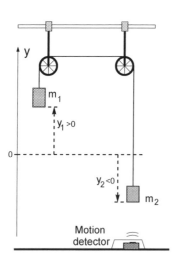

Figure 1: Atwood's machine.

Kinetic, potential, and total energy in Atwood's machine. The system you will study comprises the two masses connected by a string passing over a pair of pulleys in what is called an Atwood's machine, shown in Fig. 1. Design and carry out measurements of the total kinetic energy K, total gravitational potential energy U, and total mechanical energy E of the "system" consisting of masses m_1 and m_2 in Atwood's machine as functions of time when the masses are released from rest. You will use the ultrasonic motion detector and *Logger Pro*. The *Logger Pro* file "atwood2.cmbl" is configured to collect data from the motion detector and provides columns for distance and velocity of one of the masses.

You are to create columns to calculate all other quantities you need. Make the graph window plot K, U, and E for the system as functions of time all on the same graph. The basic parameters for the experiment and some of the considerations necessary are described below. The specific implementation will be up to you.

Use masses $m_1 = 250$ g and $m_2 = 200$ g. (Don't forget to include the mass of the hanger.) Practice letting the masses run without the motion detector present. You must step in and ensure that m_2 does not collide with the pulley and catch any masses that may fall off the hanger. Once you have the rhythm down, put the motion detector under the lighter mass, m_2, and put protective basket in place over the motion detector. The careful alignment of the pulleys in a single plane is important to minimizing friction's effects and achieving reliable data. To avoid spurious ultrasound reflections from the table, the cross bar holding the pulleys can be rotated so it is perpendicular to the table edge and the motion detector is on the floor, but a couple feet out away from the table. Once again careful positioning will allow the motion detector to track the position of m_2 accurately through out its full range of travel up. You should achieve accurate tracking for *at least 1 m* of travel.

Think carefully about how to calculate the needed quantities. Plan out your experiment and decide what additional calculations need to be performed by *Logger Pro*. You need to consider the following questions and document your choices and assumptions in your lab notebook.

- Given the position of one mass, how can you determine the position of the other? What will you choose as the origin of your coordinate system? (The "Zero" button provided by *Logger Pro* may be useful. The most convenient way to select your origin may be to adjust the masses to be at equal heights and use the zeroing function to choose this as your origin. How will y_1 be related to y_2 then?

- How are the velocities of the two masses related? What determines the total energy of the system at the start of each run?

Demonstrate and document the reproducibility of your measurements. Include a graph with all three (K, U, E) energies of the system as functions of time showing your results. Discuss how the behavior of K and U support the principle of conservation of energy.

The total mechanical energy is unlikely to be perfectly constant in your measurements. Plot and print E as a function of position. Assuming this (small) change in total energy is the consequence of a non-conservative force, $\Delta E = W_{nc}$, estimate the average force acting. (How can a non-conservative force be extracted from a graph of $E(y)$?) Is the sign of ΔE consistent with a frictional force, most likely in the pulleys?

Experimental test of friction effects. Experimentally test the effects of friction by running the Atwood's machine with equal masses, $m_1 = m_2 = 250.\,$g, and making measurements with the motion detector again, except you'll need to start the system an initial pull yourself. What should happen to the speed of the masses in the absence of friction? What is observed?

With equal masses and some initial speed, decide how your measurements can be used to determine the fraction of energy lost due to friction and to estimate the frictional forces in this case.

Kinetic energy of the pulleys. The pulleys themselves gain some kinetic energy in the original Atwood's machine measurements with $m_1 > m_2$. This contribution to the total energy has been neglected. While you may not know exactly how to express kinetic energy of the pulley as it rotates, the speed of the outer edge of the pulley ought to be the same as the speed of the string, provided the string doesn't slip relative to the pulley surface. Make a plausible guessimate at the mass of the pulleys and estimate the kinetic energy they acquire. Might this omission account for the apparent ΔE observed in the original measurements? Which effect, pulley K.E. or friction, seems to be more important? How do you reach your conclusion?

accurate from 0cm → 40cm (handwritten)

Before you leave: Be sure to rotate the Atwood's machine cross bar back parallel to the table edge so the next group of students don't bang their heads on it coming onto lab. Please clean up your table, leaving the equipment neatly organized and free of forgotten paper, scraps, calculators, etc.

Logger Pro tips: The "Zero" button provided by *Logger Pro* measures and stores the current distance to the object. Then *Logger Pro* automatically subtracts this distance from subsequent measurements so that distances are reported relative to this zero point. The zeroing function is also available from the menu bar via "Experiment - Zero...").

To create new columns of data calculated from sensor data collected, use "Data - New Calculated Column" and fill in the information for long and short versions of the column name, the units, and enter your formula. Remember, when using another column in the calculation the column name must be in quotes. (Or pick it out of the drop down list available.) You can also include defined user parameters or constants (like masses) in formulas.

8—Collisions and conservation laws

Object

To test the conservation of linear momentum in collisions on an air-track and to investigate kinetic energy changes in collisions.

Background

The linear momentum $\vec{p}$ of an object is determined by its mass and velocity: $\vec{p} = m\vec{v}$. Note that linear momentum is a *vector* quantity. If the mass of the object is constant, any changes in linear momentum can be related back to forces on the object very simply:

$$\frac{d\vec{p}}{dt} = \frac{d(m\vec{v})}{dt} = m\frac{d\vec{v}}{dt} = m\vec{a} = \vec{F}. \qquad (1)$$

The change in momentum, or impulse, due to a force on an object can be found by integrating the force with respect to time:

$$\Delta\vec{p} = \int d\vec{p} = \int \vec{F}dt. \qquad (2)$$

During a collision of two masses, m_1 and m_2, each object exerts a (time-varying) force on the other. If there are no net external forces (or the collision occurs quickly enough that the effects of external forces can be neglected), only these internal forces between objects are important. Since each object experiences a force, this suggests each object must experience a change in momentum. Newton's third law further requires that the force on object 1 due to object 2 ($\vec{F}_{12}$) must be equal in magnitude and opposite in direction to the force on 2 due to 1 ($\vec{F}_{21} = -\vec{F}_{12}$). Since the forces at every instant have equal magnitudes and opposite directions, the change in linear momentum of object 1 must be equal and opposite the change in 2's momentum:

$$\Delta\vec{p}_2 = \int \vec{F}_{21}dt = -\int \vec{F}_{12}dt = -\Delta\vec{p}_1. \qquad (3)$$

The total linear momentum, $\vec{p_1} + \vec{p_2}$, is unchanged by the collision since $\Delta\vec{p_1} + \Delta\vec{p_2} = 0$. The total linear momentum after the collision is the same as it was before the collision. The momentum is said to be *conserved*:

$$\vec{p}_{1f} + \vec{p}_{2f} = \vec{p}_{1i} + \vec{p}_{2i}.$$

This assumes no net external forces are acting on the system: the only relevant forces acting during the collision are the forces between the objects.

A collision is called elastic if the total kinetic energy is the same before and after:

$$\frac{1}{2}m_1 v_{1f}^2 + \frac{1}{2}m_2 v_{2f}^2 = \frac{1}{2}m_1 v_{1i}^2 + \frac{1}{2}m_2 v_{2i}^2.$$

If the kinetic energy is not conserved, $K_f \neq K_i$, the collision is inelastic. A perfectly (or totally) inelastic collision is one in which the objects move together as a single unit after the collision.

Procedure

The equipment for this lab includes an air-track, gliders, and photogates for timing. The photogates can be monitored by computer. By measuring the time each gate is blocked by a glider passing through, the glider's speed through that photogate can be calculated given its length. This can be done directly with the computer.

The *Logger Pro* file "collisions.cmbl" is configured to monitor two photogates and report the speeds of the glider passing through each. The length used to calculate speeds is set under "Setup - Data Collection" where photogate timing mode should be selected, and "collision timing" should be selected under "Sampling." The length used must be the same for both gliders.

The speeds reported do not indicate the direction of motion. Since momentum is a vector quantity, direction is crucial. It is up to you to add the appropriate sign to each speed the timer program reports. You must also sort out which readings correspond to the initial and final speeds for each glider. Decide how to place the photogates so that you can record all necessary initial and final velocities of both gliders in the collision scenarios described below. Plan ahead so that you can record your data in neat and thoughtfully organized tables in your notebook.

TREAT THE GLIDERS AND AIRTRACK WITH GREAT CARE. DO NOT
DROP OR HANDLE THE GLIDERS ROUGHLY. IF BENT EVEN
SLIGHTLY, THEY BECOME USELESS. PUSH THE GLIDERS GENTLY!
THEY SHOULD BE MOVING AT LOW SPEEDS, AND A COLLISION
SHOULD NOT MAKE THE GLIDERS BOUNCE AGAINST THE AIRTRACK
ITSELF DURING THE COLLISION.

Once you are ready to make the measurements outlined below, start the air-track blower. Allow a few minutes for the air-track to warm up. *Leave it on until you have completed all your measurements.* While it warms up, level the air-track carefully. Test for levelness by making sure the gliders remain nearly stationary when placed at rest on the air-track between the photogates.

Single glider—no collision. Measure the speed of one glider passing through the two photogates as it travels freely along the air-track. Measure with the glider traveling in both directions along the air-track. Make at least two trials in both directions. Keep glider speeds ≤ 30 cm/s.

The readings from the two photogates may not be identical on a single pass due to slight differences in the photogates, residual tilt or unevenness in the air-track, and small drag forces on the glider. These measurements provide you with a gauge of how significant these effects are. You will need to take them into consideration when examining momentum and energy conservation in collisions.

Inelastic collisions. Explore collisions between two gliders whenre they stick together after the collision. Make measurements of three combinations of gliders. Carry out at least two trials for each combination. Record all the relevant raw data in a table in your notebook.

- Equal glider masses ($m_1 \approx m_2$): $m_1 \longrightarrow m_2$, with m_2 initially at rest;

- Heavy glider into light glider (add ≈ 100 g to original m_1 or use a larger glider if available): $m_1 \longrightarrow m_2$, with m_2 initially at rest;

- Light glider into heavy glider: $m_2 \longrightarrow m_1$, with m_1 initially at rest.

Mass 1	Mass 2	v_{1i}	v_{2i}	v_{1f}	v_{2f}
...	...	...	...	...	...

(Nearly) elastic collisions. Repeat the same combinations (at least two trials of each) using the bumper ends of the gliders and gather the same kinds of information. Again collect all the data in a table in your lab notebook.

Analysis

Computers can reduce much of the tedium of calculating by hand the momenta and kinetic energies of individual gliders as well as totals and changes. Using *Logger Pro* the raw measured data can be entered into a data table and then columns created that automatically compute the other needed quantities. A file, "collisions_analysis.cmbl," provides a beginning data table template for you to extend in completing the calculations. You will need to enter your mass and velocity data and create additional calculated columns to complete the analysis. Refer to notes provided in the file for information.

Momentum analysis. To test momentum conservation, construct additional columns for the:
- initial and final momenta the gliders individually ($\vec{p}_{1i}$, $\vec{p}_{2i}$, $\vec{p}_{1f}$, $\vec{p}_{2f}$),
- total initial and final momenta $\vec{p}_i$, $\vec{p}_f$,
- change in total momentum, $\Delta\vec{p} = \vec{p}_f - \vec{p}_i$,
- percentage change in momentum, $\frac{\vec{p}_f - \vec{p}_i}{|\vec{p}_i|} \times 100\%$.

Each row of the table should present the results of one of your collisions, beginning with the trials of a single glider traveling along the air-track alone.

Energy analysis. To examine kinetic energy conservation in the collisions, construct a table showing the total kinetic energy before and after, and the percentage change in total kinetic energy, $\frac{K_f - K_i}{K_i} \times 100\%$. (It may be a convenient but not essential intermediate step to create columns calculating the initial and final kinetic energies of the individual gliders, K_{1i}, K_{2i}, K_{1f}, K_{2f}.)

Include copies of the tables showing the momentum and energy analyses in your notebook.

Before you leave: Please be sure to save your analysis file and take it with you. Leave your table clean and equipment ready for the next lab section.

Post-lab Questions

1. Roughly what percentage changes in momentum were observed for the single glider runs through two photogates? Was there a pattern as to increases or decreases? Did the direction of travel seem to matter? If so, why might that be?

2. Given the experimental uncertainties as typified by the measurements of a single glider traveling successively through two photogates, to what extent do the results support the conservation of momentum in your two groups of collisions? To what extent do conditions of the experiment meet the conditions required for momentum to be conserved?

3. Which collisions most nearly conserve kinetic energy?

4. For the inelastic collisions, assume momentum conservation is true and derive an expression of for $\Delta K/K_i$ that involves only the glider masses. (Make use of the fact that m_2 is initially at rest in your collisions to simplify the algebra.) Compare the fractional changes in kinetic energy you calculated from your inelastic collisions to the value predicted by this result.

5. A measure of how a elastic a collision was can be described by the elasticity or coefficient of restitution:

$$e = \frac{|\vec{v}_{1f} - \vec{v}_{2f}|}{|\vec{v}_{1i} - \vec{v}_{2i}|}.$$

An elastic collision is characterized by $e = 1$. Find e for three of your totally inelastic (sticking) collisions (very easy!) and three of your nearly elastic collisions (try one of each glider configuration), and discuss the differences. How close to $e = 1$ are your "elastic" collisions?

6. What do you expect a graph of total final momentum *vs.* total initial momentum for all your collisions on a single graph to look like? Why? Make such a plot. How does your graph compare with your expectations?

7. Make a similar graph for the kinetic energies and comment on the result.

Logger Pro tips: If the data table becomes inconveniently large, you can temporarily hide columns in *Logger Pro*. Double-click at the upper-left of the table (or click once on the table to select it and then from the menu bar select "Options – Table Options") and uncheck columns you want to hide. These can be customized on each page of the *Logger Pro* file. Once you have the columns you want to display, you can print that portion using 'File - Print Data Table.' Remember that printing in "landscape" orientation may give a better layout on the printed page.

9—Ballistic pendulum

Object

Apply conservation principles to the motion of a ballistic pendulum to determine the velocity of a projectile, and to predict the range of that projectile when launched horizontally from a tabletop.

Background

The ballistic pendulum is a device used to catch a ball fired from a spring-powered gun. Just after the ball (with mass m) is launched, it is caught by the pendulum bob (mass M), and the combination moves together. The apparatus includes a mechanism that records the ball/bob combination at or very near its maximum height after the collision. Since the ball and the pendulum move off together after the collision, it is a perfectly inelastic collision. An analysis of the collision and subsequent swinging motion should use energy and momentum considerations. In analyzing the motion, it is useful to divide it into three pieces or processes: (1) the ball's approach just before it strikes the bob, (2) the collision and capture of the ball in the bob, and (3) the swinging motion after the collision to the final state at the top of the pendulum swing. Consider the following questions to deduce the initial speed of the ball, v_o.

- Is momentum conserved during the collision? Is it conserved as the pendulum swings away and up?

- Is kinetic energy conserved during the collision? Is it conserved during the swinging motion?

- During any portion of the motion is the total mechanical energy (kinetic plus potential, $E = K + U$) conserved?

The effective center of mass of the ball and pendulum bob combination or the effective length of the combination will be marked or given. This

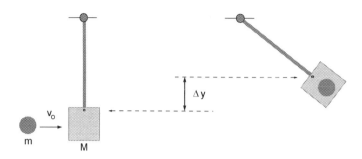

Figure 1: Ballistic pendulum.

can be useful in evaluating changes in gravitational potential energy when treating the ball/bob system as a point mass. If you regard the pendulum as a point mass, its kinetic energy becomes $\frac{1}{2}(M + m)v^2$ and potential energy is simply $(M + m)gy$. (Although not all parts of the pendulum move with the same speed after the collision—the top of the support rod for example moves more slowly than the bottom—this effective center of mass accounts partly for such effects.)

Procedure

Develop a way to figure the ball's initial velocity based on the height achieved by the ballistic pendulum and energy and momentum concepts mentioned above. Fire the ball 10 times into the pendulum. Find the average initial velocity of the ball and make an estimate of the uncertainty in your value.

SAFETY FIRST!
DO NOT LEAVE YOUR LAUNCHER LOADED
AND ARMED UNLESS YOU ARE ACTIVELY
MAKING A TRIAL!

Given this initial velocity, use kinematics of projectile motion to predict where the ball will land on the floor when it is fired from the tabletop with the pendulum removed. Include an estimate of the uncertainty in this predicted range. Tape a piece of white paper on the floor and mark the expected landing spot with an "X". You can use a plumb bob and meter stick to mark this location carefully. Also mark on the paper your estimated uncertainty in the landing point. When you are ready, ask your lab instructor to observe your test of your prediction. Cover the white paper with a piece of carbon paper (don't tape it) and carry out five test shots.

MAKE SURE THE FIRING RANGE IS CLEAR BEFORE
LAUNCHING THE BALL!

Measure the actual range, report an average and standard deviation for your trials. Infer from the measured range the initial velocity of the ball and the uncertainty. If there are large discrepancies between predicted and measured ranges (or in the velocities) obtained from the pendulum and projectile measurements, review your analysis and calculations. Discuss any other smaller discrepancies. Do they suggest some systematic error? If so, can you suggest possible sources for systematic errors that might account for the results observed? Be specific as to how your proposed source of error would produce the any observed systematic difference.

Make sure the *principles and method* of your analysis are clearly presented in your lab notebook along with the data and calculations.

Before you leave. Please be sure to remove all tape and paper from the floor and leave your lab table clean and organized for the next lab section.

10—Rotational dynamics

Object

To study the relationship between torque and angular acceleration, to measure the moment of inertia of a ring, and to find the total energy of a system with rotational and linear motion.

Background

Newton's second law analog for a rotating system is $\tau = I\alpha$. τ is the net torque acting on the system whose rotational inertia (or moment of inertia) is I. α is the resulting angular acceleration. You will test this relationship by applying different torques and measuring the angular acceleration. The ratio $\frac{\tau}{\alpha}$ should be the same for each case.

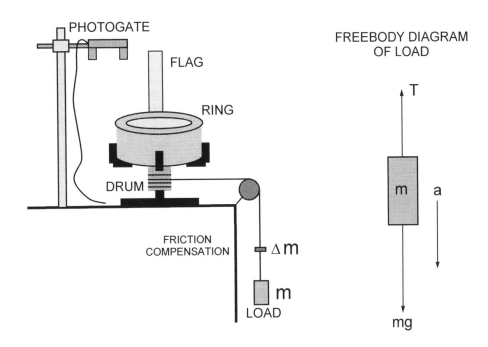

Our system consists of a heavy steel ring mounted on a light metal frame, which rotates on low friction bearings. A torque is applied to the system by a string wound on a small drum mounted on the rotating frame. The tension in the string is provided by a small falling mass, m. As the system accelerates the string unwinds and the mass falls.

The relationship between the linear acceleration, a, of the falling mass and the angular acceleration, α, of the rotating system is given by $a = \alpha r_d$, where r_d is the radius of the *drum*, a part of the rotating frame. Do not confuse this radius with the radius of the heavy ring, which we will denote by R.

The tension in the string is not simply the weight (mg) of the mass. The mass is accelerating. Using the free-body diagram of the falling mass and Newton's second law gives $mg - T = ma$, so that the tension in the string is $T = m(g - a)$, taking downward acceleration as positive. Expressed in terms of the angular acceleration,

$$T = m(g - \alpha r_d).$$

The torque on the rotating platform due the string is then the lever arm or distance from the rotation axis, r_d, times the tension:

$$\begin{aligned} \tau &= r_d T \\ &= r_d m(g - \alpha r_d). \end{aligned}$$

From the rotational analog of Newton's 2nd law, $\tau = I\alpha$, the ratio of torque to angular acceleration

$$\frac{\tau}{\alpha} = I = \frac{m r_d (g - r_d \alpha)}{\alpha},$$

This can be re-written as

$$I = \frac{mg r_d}{\alpha} - m r_d^2. \tag{1}$$

The strategy to test the second law for rotation is to find α experimentally and use Eq. 1 to get the moment of inertia, I. If this is practically constant for different m's, then angular acceleration is proportional to torque, and this form of the law is correct. You will later check its validity by calculating the moment of inertia by another method and comparing the values of I found.

For a given load mass m, the angular acceleration of the ring is expected to be constant as the mass falls. Equations describing the motion are analogous to those familiar from 1-D motion:

$$\omega(t) = \alpha t + \omega_o \tag{2}$$

$$\theta(t) - \theta_o = \frac{1}{2}\alpha t^2 + \omega_o t \tag{3}$$

$$\omega^2 = 2\alpha(\theta - \theta_o) + \omega_o^2. \tag{4}$$

The last of these, Eq. 4, is the one you will use to measure α. A graph of ω^2 as a function of θ is expected to produce a straight line. The slope of that line is *twice* the angular acceleration.

Procedure

Preliminaries. Measure the radius of the drum, r_d, that the string winds around. Measure the mass, M, of the ring. It may have a value written on it, but measure it using the one large capacity balance in the lab. Measure the inner (D_i) and outer (D_o) diameters of your ring and figure the outer radius R where the flag is mounted.

Timing system. The motion of the ring will be measured with a photogate and computer. The system records the time required for a flag of width W attached with tape to the *outside* of the ring to pass through the photogate on each revolution. (Don't tape the flag to the retaining arms of the frame (wrong radius!), or place the ring on the frame so you have a layer of tape forced between the ring and the retaining arms of the frame.) The *Logger Pro* experiment file "rotational_dynamics.cmbl" contains the beginnings of the data collection and analysis process. This file is configured to keep a count of the number of revolutions or passages of the flag through the photogate (starting at 0 on the first pass), report for each passage the time required for the flag to pass through (Δt), and calculate the flag's speed, based on a user-provided width of the flag, $v_f = \frac{W}{\Delta t}$.

Measure the width of your flag and enter an accurate value into the experiment under "Experiment - Setup Sensors - LabPro 1 - Dig/Sonic Phototgate - Set distance or length." Test out the operation of the basic file.

From the number of revolutions completed you can express the angular position θ. A calculated column gives the angular position in radians with the definition of $\theta = 0$ at the first pass through the gate. (θ increases by 2π for each revolution, so $\theta = 2\pi \times$ # of revs completed.) Using the ring's radius R and v_f, make a column for the angular velocity ω at each passage through the photogate. (How are ω, v_f, and R related?)

Friction compensation. This system does have friction, primarily in the bearings that support the rotating frame. It produces a frictional torque on the system. You can approximately compensate for friction by a small additional mass on the string that produces an equal and opposite torque on the system.[1] This is found by trial and error. After the apparatus has been set

---

[1] The frictional torque is assumed to be the same at all speeds. When the system is accelerating the friction compensation mass will also contribute a very small additional torque on the system, but this term produces an error that is much smaller than the other sources of errors in this experiment and will be neglected.

up, wind the string carefully on the drum with the turns close together but not overlapping. The string should come off the drum as nearly horizontal as possible. Place a small mass (e.g. small washers or paper clips) on the end of the string and give the system a modest angular velocity with a gentle push, allowing the mass to fall slowly. If Δt, the time to pass through the gate, does not remain nearly constant on successive passes, add or subtract mass until the system is closest to maintaining a constant angular velocity.

Dynamic measurement of α and I. According to Eq. 4, ω^2 is a linear function of θ. A plot of ω^2 as a function of θ should be a straight line of slope 2α. Make the necessary additions to the data table and changes to the graph window to produce such a graph. (Plot $Y = \omega^2$ and $X = \theta$) .

To start a data run, wind the string carefully on the drum with the friction compensation mass and an additional load mass of $m = 20\,\mathrm{g}$ on the end of the string. With the load mass near the top, position the ring so that the flag is about to enter the photogate. Start data collection in *Logger Pro*, then release the ring.

What does the resulting graph tell you? Does it show that the angular acceleration is constant? Use the linear fit tool to figure out the angular acceleration. Repeat with the same load mass at least twice more and record the angular acceleration for each. Store (Ctrl-L) at least one data run for later comparisons and printing. Compute the average α. Calculate I using Eq. 1. Pay attention to units; I should be reported in $\mathrm{kg\,m^2}$.

Repeat the procedure with load masses of $m = 50\,\mathrm{g}$ and $100\,\mathrm{g}$. Save all the data to a file for later review if needed.

Print a graph that shows at least one run from each load mass for your lab notebook. Include the corresponding data tables.

Are the values of I consistent? What is the average I? What is your estimate of uncertainty in I based on these measurements?

Energy measurements. Clear all stored data in *Logger Pro* after saving it for later review. Add a column to the data table in *Logger Pro* to calculate the total kinetic energy of the system: rotating system *and* falling mass:

$$K = \frac{1}{2}I\omega^2 + \frac{1}{2}mv^2.$$

- Use I found for the $m = 100\,\mathrm{g}$ load. How is the speed of m related to ω? Pay close attention to the units! Energies should be reported in joules.
- Add another column that calculates the potential energy of the system — the gravitational potential energy of the falling mass, $U = mgy$. Use $m = 100\,\mathrm{g} = 0.100\,\mathrm{kg}$. How is m's position related to θ? Let $U = 0$ at the top.
- Add one more column that finds the total energy $E = K + U$.

• Modify the graph to plot K, U, and E as functions of θ.
Before collecting data, think about what these plots should look like. What do you expect to see for K and U individually? How should the total energy behave? Try two or three runs using $m = 100\,\mathrm{g}$; leave the friction compensation mass in place.

Print out the graph for your notebook and save your data.. Include a data table from one run in your notebook. Does the graph demonstrate the conservation of energy? Discuss how. Discuss probable reasons for any discrepancies.

Moment of inertia of frame. The experimental value of I should be slightly greater than a predicted value of I_{ring} since the experimental value also includes the rotational inertia of the frame: $I = I_{ring} + I_{fr}$. The comparatively small rotational inertia of the frame, I_{fr}, makes its precise measurement difficult, but even a rough value that has a relatively high percentage uncertainty can improve the value of the measured I of the ring without adding much to the uncertainty in the determination of I_{ring}.

With the ring removed from the frame, compensate for friction, then make a determination of I_{fr} as before, using a single load of 10 g. Subtract this value of I_{fr} from I of the whole system to get improved measures of I_{ring}.

Calculated moment of inertia of the ring. To check on the experimental values of I_{ring} for all 3 loads as found above, calculate the predicted moment of inertia of the heavy steel ring from the formula

$$I_{ring} = \frac{M}{8}\left(D_i^2 + D_o^2\right).$$

Compare this calculated value to the values for I_{ring} obtained from the earlier measurements, including the corrections made for the contribution from the frame.

Before you leave. Organize the equipment neatly on the table, and leave the table free of random paper, etc. Be sure the ring is re-mounted on the frame and the flag is properly attached for the next group to begin work.

11—Archimedes' principle

Object

To design and carry out a measurement of specific gravity using Archimedes' principle and Hooke's law with a limited set of measuring tools.

Background

Archimedes' principle asserts that a solid object in water (or any fluid) experiences a buoyant (upward) force exerted by the water pressure that is equal to the weight of the water (or fluid) the solid displaces: $F_B = \rho_w V_x g$. V_x is the volume of water displaced by the solid.

The specific gravity of a material is the ratio of its density to the density of water, ρ_x/ρ_w. For materials with a specific gravity less than 1 ($\rho_x < \rho_w$), the object will float in water, since the buoyant force will equal the weight of the object $m_x g$ before it is completely submerged. If the specific gravity is greater than 1 ($\rho_x > \rho_w$), the object will sink, but its apparent weight in the water will be reduced by the buoyant force, $W' = m_x g - F_B$.

Since the density of water is very nearly 1.00 g/cm^3 (=1000 kg/m^3 = 1 Mg/m^3 and it varies slightly with temperature), the specific gravity also corresponds closely to the numerical value of the density of the object when ρ_x is measured in g/cm^3 or Mg/m^3.

Procedure

Specific gravity of a solid. You are provided with water, a spring, a beaker, some string, a ruler, and an irregularly shaped object whose specific gravity you are to determine. Using only these tools (no balances or volume measurements), Archimedes' principle, and Hooke's law, invent and execute a method to measure ρ_x/ρ_w for the sample provided. Your lab team should discuss options and possible methods and settle on a single satisfactory approach. Make revisions to your method as necessary after trying it. (Note: the brass weight attached to end of the spring is there merely to ensure that the spring is always slightly extended and will then follow Hooke's law. Your unknown sample should hang below it on a string.)

Document your method by explaining the reasoning behind it at the start of your report, before making measurements. Include freebody diagrams and develop the necessary equations in *symbolic form* using standard notation, e.g. k for spring constant, ρ_w for density of water, ρ_x for the unknown density, V_x for the volume of the unknown sample, etc.

Clearly present your measurements and final calculations of ρ_x/ρ_w. Make estimates of uncertainty in each of your measurements and estimate the probable error in your value for the specific gravity of the unknown.

Specific gravity of a liquid. Incorporate your previous results in determining the specific gravity of the additional liquid provided.

12—Latent heat of fusion

Object

To measure the latent heat of fusion for H_2O.

Background

Phase transitions—changes in a substance from solid to liquid (fusion or melting) or liquid to vapor (vaporization)—require the addition of energy. During the phase change energy added does not produce a temperature change in the substance until the substance has completely melted or vaporized. In this lab the thermal energy required to melt ice will be determined. The latent heat of fusion, L_f, is the energy required to melt a unit mass of a substance, and we will measure it in J/g. To melt a mass m of a material that is already at the melting temperature requires heat transfer of $\Delta Q = mL_f$.

A known mass of ice at $0\,°C$ added to warm water (at T_1) will melt and lower the temperature of the water. Given enough warm water the final temperature will be between $0\,°C$ and T_1. The heat (thermal energy) lost by the warm water ($\Delta Q = mc\Delta T$) is absorbed by the ice causing it to melt. The water/ice system can also exchange energy with the surroundings during the melting. To reduce this exchange a calorimeter is used. A calorimeter is a thermally insulating container that ideally reduces energy transfer to the surroundings to zero. In practice, a good calorimeter reduces the heat exchanged with the surroundings to negligible levels. The double-walled calorimeter here is an aluminum cup that holds the water and ice, which sits inside a second larger cup on an insulating plastic ring. A cover closes the system. If the heat transfer between the inner cup with ice–water mix and the surroundings is negligible, then we can apply the conservation of energy to the melting process, but we must also account for the heat capacity of the inner aluminum cup as part of our system. One way to express this idea that the total change in (thermal) energy must be zero ($\sum \Delta Q = 0$ in terms of gains and losses is:

(energy lost by warm water) + (energy lost by calorimeter cup)	$=$	(energy absorbed to melt ice) + (energy absorbed to warm liquid to the final temperature).

Procedure

Preliminary analysis. Let T_1 and T_2 be the initial and final temperatures (measured in °C) of the water. m_w, m_i, and m_c are the masses of the initial warm water, ice, and calorimeter cup (including the wire stirrer), respectively. The specific heat of water is denoted as c_w, and for the aluminum cup and stirrer it is c_{Al}. L_f is the latent heat of fusion for ice. Express the heat balance statement above in mathematical form in terms of these quantities. Assume the ice is initially at $0\,°C$.

Solve your heat balance equation for L_f in terms of the other quantities.

Experimental procedure. To ensure the ice is initially at $T = 0°C$, put 3-4 ice cubes in the small plastic insulating container with a small amount of water and allow the mixture to come to equilibrium while you make other preparations.

Record room temperature. (The *Logger Pro* file "thermometer.cmbl" is configured to display and graph the temperature.) Weigh the empty calorimeter cup and stirrer. Make a very quick and rough measurement of the mass of a single ice cube. Fill the cup about 2/3 full with warm water 6 to 10°C above room temperature. Determine the mass of the water added.

Place the calorimeter cup and contents into the outer can of the calorimeter, supported by the plastic ring. Cover, insert the thermometer, and stir gentlyfor at least a minute, until a stable initial temperature is achieved. Record this starting value, T_1.

Place a dried (use a paper towel) ice cube or two into the calorimeter. (30-40 g is appropriate. Determine the precise ice mass added later!) Avoid splashing. Stir gently but frequently (just enough to ensure good mixing) until a minimum temperature is reached. Record this as T_2.

Find the mass of the ice added.

The specific heat of (liquid) water is $c_w = 4.18\,\text{J/g°C}$, while for the aluminum cup and stirrer $c_{Al} = 0.900\,\text{J/g°C}$. Calculate the latent heat of fusion for ice from your measurements. (A note on units: 1 J/g°C = 1000 J / kg·K, since a change of 1 °C is a change of 1 K.)

2nd trial. Repeat the experiment with a similar amount of ice and warm water. This time, use your prior experience to adjust the starting water temperature so that your expected final temperature, T_2 is about as far below room temperature as your T_1 is above room temperature. (For example if in your first trial your warm water was 6° above room temperature and the final temperature was 10° below room temperature, start with water about 2° degrees warmer. This reduces the effects of heat exchange with the surroundings during the experiment. Keep track of the time that elapses during the melting. Calculate L_f again.

What is your average value of L_f? Given your two measurements, how many significant figures do think it is reasonable to report from this experiment? Is your value consistent with the value reported in textbooks?

Interaction with surroundings. Arranging for initial and final temperatures to be equally far from room temperature reduces the effects of heat exchange with the surroundings during the experiment. Why will this choice of T_1 have this benefit?

The influence of heat transfer from the surroundings can be gaged roughly. Begin the measurement cycle again with warm water, but don't add ice. Instead, stir occasionally and simply monitor the temperature change in the calorimeter that occurs during the amount of time that elapsed in the second trial. Based on the temperature change observed, calculate how much heat was transferred to the surroundings during the measurement. Compare this to the amount of heat that was transferred from warm water and calorimeter to melt and warm the ice during your second trial.

Is the calorimeter set-up used in the experiment effective enough at thermally isolating the inner cup from its surroundings?

Thermometer omission. The thermometer is not included in the heat capacity of the calorimeter. Assuming it is made mostly of stainless steel, make a rough estimate of its heat capacity. Compare this to the heat capacity of the inner calorimeter cup and the heat capacity of the original warm water. Is this a significant omission? Why or why not?

13—Electrostatics

Object

To make qualitative observations of electrostatic phenomena and interpret the phenomena in terms of the behavior of electric charge in various situations. To apply Coulomb's law to estimate the charge on objects.

Background

A neutral atom of a substance contains equal amounts of positive and negative charge. The positive charge resides in the nucleus, where each proton carries a charge of $+1.602 \times 10^{-19}$ coulombs. The negative charge is provided by an equal number of electrons associated with and surrounding the nucleus, each carrying a charge of -1.602×10^{-19} C. Macroscopically-sized samples of everyday materials usually contain very nearly equal numbers of positive and negative charges. When some dissimilar materials are rubbed together, some charges are transferred from one material to the other, usually as charged pieces of molecules broken off by the frictional forces of rubbing, leaving each object with a small but discernible net charge. For example, when a glass rod is rubbed with silk, the rod usually ends up with a positive charge.

The magnitude of the electric forces charged objects exert on one another is given by Coulomb's law

$$F = k \frac{q_1 q_2}{r^2}.$$

The objects carry electric charges q_1 and q_2 and are a distance r apart. The direction of the forces depends on the sign of the charges. Like charges exert repulsive forces on each other. Charges carrying opposite sign exert attractive forces on each other. The constant $k = \frac{1}{4\pi\epsilon_o} = 8.99 \times 10^9$ Nm^2/C^2.

Materials can be cast into two rough electrical categories: insulators and conductors. The atomic or molecular structure of the material determines whether some charges are free to move (making a conductor, such as metallic materials), or largely fixed in place (an insulator). A conductor lets charges (electrons in most solid conductors) move freely throughout and allows the transfer of charge through it from one object to another. Insulators generally prevent charges from moving far; electrons are typically localized to a particular atom. A charge put on an insulator stays put, unlike charges placed on a conductor. Even in insulators, the charges in the material can respond to electric forces in a way that causes small displacement of opposite charges in opposite directions, leading to electric polarization.

You will carry out a series of tests. In many cases you are asked to observe various phenomena and develop a qualitative explanation of how the electrical nature of matter at the microscopic level accounts for the observed behavior.

DUE TO THE LARGE ELECTRIC FIELDS PRESENT IN PARTS OF THIS EXPERIMENT AND POTENTIAL FOR LARGE ELECTROSTATIC DISCHARGES, SENSITIVE ELECTRONIC EQUIPMENT, INCLUDING CALCULATORS AND COMPUTERS, SHOULD BE KEPT AWAY FROM THESE HAZARDS, ESPECIALLY THE VAN DE GRAAFF GENERATOR.

Procedure

The success of electrostatics experiments depends on weather and the condition of the materials. Many of the following observations can be challenging in humid conditions. One action that can help you obtain reliable results is to clean the various rods used with a cloth moistened with a little rubbing alcohol to make sure they are free of excess moisture or grease from frequent handling. Once cleaned, be careful to handle them consistent at one end. A second good practice is handle the fabrics as little as possible and to lay them flat to dry between use. These steps will help make it easier to generate more electrostatic charge.

Figure 1: Electroscope.

Frictional charging and the electroscope. Look carefully at the construction of the electroscope. It consists of a large metal ring surrounding a fixed vertical post with a zag in the middle. A metal needle is pivoted in the middle of the post. The needle and the post are all metal (conductors) but are separated from the surrounding ring by a plastic insulator at the top of the ring. A metal disk at the top is also connected to the needle/post combination but also insulated from the surrounding ring.

Several different rods and patches of materials or fabrics are provided. Rub various combinations of rod and material together. Three to five quick strokes is sufficient more just tends to transfer moisture and oils from hands to the fabric from your hands. Bring each rod close to the top disk on the electroscope *without making contact or hearing a small snap of a tiny spark.* Which combination of rod/fabric provides the strongest response of the electroscope when the rod is brought near the electroscope? Sketch the electroscope and how you think charges are arranged on the electroscope and rod when the rod is near. Does the electroscope carry a net charge when the rod is near? After the rod is moved away?

With the combination providing the strongest response, charge the rod by rubbing again and allow it to touch the electroscope's disk and remove. What happens to the electroscope? Repeat the charging and contact process a few times in succession. What happens? Make a simple sketch of how you think electric charge is now distributed on the electroscope that accounts for the deflection.

Charge up the electroscope with this strongest combination, then try other combinations of rods and materials. Bring the rod for each new combination close to the charged electroscope disk (without touching again), then remove the rod. Which combination of rod and fabric has the strongest *opposite* charge? *Explain* the reasoning by which you know the second rod carries the opposite charge. Sketch how you think the charge is distributed on the electroscope and nearby rod for one case each where the charged electroscope deflection increases with the additional rod's approach and decreases with the rod's approach.

Taking the charge on a glass rod rubbed with silk to be positive, which combination of rod and fabric produced the strongest positive charge? Which combination produced the strongest negative charge?

Bring a rod you believe to be *uncharged* near the charged electroscope. Does anything happen? If so construct an explanation, using simple diagrams again.

Charging by induction. While touching the edge of the electroscope disk with your finger, bring a strongly charged rod near the disk without touching. Then remove first your finger (from the disk, not from your hand!), and then the rod. What happens? Determine the sign (explain how you do this) of the charge on the electroscope. Explain how it got there.

Connect a metal wire to the disk with an alligator clip with a rubber boot on it, and connect the other end of the wire to "ground." (Connect the other end to a metal pipe at a sink or to the lab table's metal frame.) Again bring a strongly charged rod near the disk, and carefully disconnect the ground wire from the disk, touching only the plastic or rubber on the clip. Remove the rod. <u>What happens?</u> <u>What charge is on the electroscope?</u> (You might try to repeat this process with a piece of string—an insulator—replacing the wire and see if the behavior is different. If conditions are humid, the string, however, may turn out to be not a very good insulator.)

Polarization. Suspend a metal rod horizontally in a metal stirrup hanging on a string and allow it to come to rest. Strongly charge up another rod. Bring the charged rod close to the suspended rod without touching. Describe what happens. Construct an explanation. Try suspending a glass or plastic rod as well.

Bring a charged rod near a gentle stream of water from a faucet. (A nozzle that produces a narrow stream of water is best. Keep all fabric material dry and well away from the water.) Describe what happens. Does this mean water has a net electrical charge? Try the same thing with an oppositely charged rod and describe your observations.

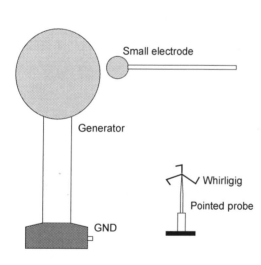

Figure 2: Van de Graaff generator and accessories.

Van de Graaff generator experiments. The Van de Graaff generator can produce significant electric potential (voltage) when it transports electric charge to the spherical dome by a rubber belt. You can remove the dome to see the belt and metal "comb" that helps transfer charge from the belt to the dome. A similar comb arrangement is at the other end of the motor driven belt, also connected to the "grounding" screw at the base of the generator. Caution: you should always unplug the generator and discharge the dome to ground with the smaller spherical electrode. "Grounded" here means connected by a wire to the grounding screw on the base of the generator) or with the adjustable U-shaped discharge electrode before making any changes or touching or working close to the generator. An unpleasant electric shock can result from even near-contact with the charged dome.

Start the generator. Holding the smaller spherical electrode by the plastic handle and no wire connecting it to ground, bring it into contact with the generator dome. (If the rod is held loosely in the hand, you can sense the forces between the dome and electrode and see the effects of attraction as you bring the electrode toward the dome and repulsion immediately after charge is transferred.) Then bring the smaller electrode into contact with an electroscope to charge the electroscope. Determine the sign of the charge taken from the Van de Graaff generator.

Ground the smaller electrode and mount horizontally, level with the center of the Van de Graaff dome about 30 cm away. Turn on the generator and determine how close the spheres must be before a discharge arc occurs periodically between the two. (You may hear and see—perhaps even feel—evidence of arcing to other places when the two spheres are far apart.) What is happening to produce this phenomenon?

Ground the metal shaft of the sharply pointed probe and hold well away from the generator. Re-start the generator and re-establish the discharge between generator dome and smaller spherical electrode. Then gradually bring the pointed probe toward the generator dome. Describe what happens to the arc. How far away is the pointed probe when the discharge to the smaller sphere ceases? What might be a practical application of this effect of a sharply pointed electrode in the presence of strong electric fields?

30 cm

Put the pointed probe on the table near the generator and place the three-armed whirligig on top. With the probe shaft grounded turn on the generator. What happens? Repeat with the ground wire removed from the shaft. What happens now? Try to explain the behavior you observe.

Coulomb's law. This can be difficult unless conditions are very dry and good static charges can be reliable transfered and retained. Suspend two pith balls from strings next to each other. Charge each by making contact with a strongly charged rod, or use the charged spherical electrode from the van de Graaff generator.) Try to charge each equally by treating each in the same way. Make repeated contacts with the rod if necessary to achieve a significant deflection from vertical. Make measurements that will allow you to use Coulomb's law to *estimate* (a precision measurement isn't expected) how much charge is on a ball, assuming equal charge on each, $q_1 = q_2$. The mass of the pith balls is roughly 25 mg. Hint: Draw a freebody diagram for one ball. If the conditions aren't right for this part (for example high humidity), you may find it difficult to deposit or keep enough charge on the pith balls to achieve a useful deflection. If so, try using one pith ball and the smaller spherical electrode charged with the generator. A plausible assumption at this stage is that if the pith ball and electrode touch, they will share charge so that the charge per unit area ($\sigma = q/A$) is about the same on each, so $q_{pith} = \frac{A_{pith}}{A_{sphere}} \cdot q_{sphere} = \left(\frac{D_{pith}}{D_{sphere}}\right)^2 \cdot q_{sphere}$, where the D's are the diameters of pithball and electrode.

Figure 3: Repulsion of charged spheres can be modeled as point charges.

~~Estimate roughly how many electrons~~ (fewer or extra) the pith ball needs to have such a charge.

Before you leave: Make sure your lab table is clean and ready for the next group. Leave equipment neatly organized for the next lab. Please don't leave random scraps of paper, calculators, water-bottles, etc., behind. Use recycling bins down the hall, not the trash, for recyclable materials.

14—Electric charges and fields

Object

(1) To explore the electric field produced by various arrangements of point electric charges and develop the ability to visualize and predict electric field patterns. (2) To build intuition about the way charges move in response to electric fields by arranging static charges so as to guide the motion of a mobile charge.

Background

Every electrically charged object creates around itself an electric field that provides the means by which it exerts forces on other charges. The basic building block is the electric field created by a point charge q:

$$\vec{E} = \frac{1}{4\pi\epsilon_o}\frac{q}{r^2}\hat{r}$$

This is the electric field created at a place (the field point) a distance r from the charge, and is a unit vector whose direction points from the charge to the field point. (The proportionality constant $1/4\pi\epsilon_o = 8.99\times10^9$ Nm^2/C^{2}2.) When multiple charges are present, the net electric field at a point in space is the superposition, or vector sum, of the fields produced by each electric charge at that location:

$$\vec{E}_{net} = \frac{1}{4\pi\epsilon_o}\frac{q_1}{r_1^2}\hat{r}_1 + \frac{1}{4\pi\epsilon_o}\frac{q_2}{r_2^2}\hat{r}_2 + \frac{1}{4\pi\epsilon_o}\frac{q_3}{r_3^2}\hat{r}_3 + \ldots$$

A charge q_0 that experiences an electric field created by other charges is subject to an electric force: $\vec{F} = q_0\vec{E}$ and will respond by accelerating according to Newton's second law: $\vec{a} = \vec{F}/m$. Depending on the sign of q_0, this acceleration will be in the direction of the net electric field (q_0 positive) or exactly opposite the field direction (q_0 negative). Note that this does not mean the velocity will always be in the direction of the field.

Two simulations developed at the University of Colorado (phet.colorado.edu) will be used: the first to study the electric fields produced by several static electric charge configurations, and the second to observe and try to guide the motion of an electric charge toward a goal in a version of "electric-field hockey."

These simulations can be found in the PHYS2016 folder on the Windows desktop in the sims sub-directory.

Procedure

I. Charges and Fields Simulation

Double-click on the charges-and-fields_en.jar Java archive file to start this simulation. It provides reservoirs of positive and negative charges (+/- 1.0 nC each, n= nano = 10^{-9}), which may be dragged onto the open working area. Before doing anything with the charges: (1) Drag the "equipotential" gadget off to the left and out of the way—we will not use it; (2) Click on the "Show E-field" box to check that box; (3) Check the "grid" box. The scale is such that each small box of the grid that appears is 10. cm on a side.

(1) Preliminaries. Drag one positive charge from the reservoir and put it in the work area. The program calculates and displays the electric field at regularly spaced points around the charge and draws an arrow to show the field direction and uses the color intensity as an indication of the electric fields strength or magnitude. (The display of these arrows is controlled by the "Show E-field" option.) Move the charge around and observe that the E-field pattern "follows" the charge.

Put the positive charge towards the left side of the working area. To measure the field more carefully, you can also drag an electric field sensor into the work area. It will display the electric field at its location as a vector, with its length now a measure of the fields magnitude. To get more quantitative measurements of the electric fields strength click the "Show numbers" box. You can see the magnitude and direction of the vector displayed at the sensor. The units displayed are V/m (volts per meter), which are the equal to newtons per coulomb (N/C), which may be more familiar to you at this stage.

- Play with these tools a bit, including exchanging a negative charge for the positive charge, so that all lab team members are comfortable working the simulation. Describe qualitatively the main difference between the electric field pattern for a positive and negative 1 nC charge.

- As a quick quantitative check, predict, using the equation for the electric field of a point charge given earlier, the electric field of a single +1.0 nC at distances of 1.0 and 3.0 m and verify that the simulation gives the same result. Show how you make this prediction in your lab notebook not just a final number!

- Put down four charges of mixed signs in any pattern you like and sketch the resulting field pattern.

(2) Field of an electric dipole. Place one positive and one negative charge on the grid, a distance d = 1 meter (2 large boxes) apart roughly as shown. This charge configuration is known as an electric dipole. Consider the center of these two to be the origin of a coordinate system for later measurements.

First, sketch in your lab notebook the resulting pattern of electric field lines. In particular, note the direction of the electric field along the horizontal axis (x-axis, along the dipole): (a) near the origin between the charges, (b) to the left of the negative charge and to the right of the positive charge. Also observe the direction of the field along the vertical (y) axis.

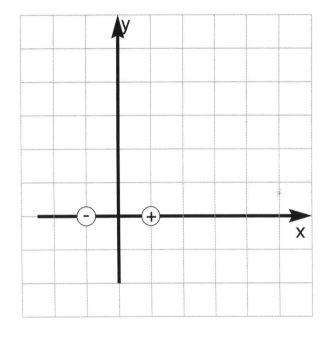

Next put a field sensor on the x-axis at $x = 2$ meters (4 large boxes from the center of the dipole. Record the strength of the electric field for the initial dipole. Put a second field sensor at 2 m above the dipole on the y-axis. Then in successive steps, shrink the dipole by moving the charges closer together in steps of 10 cm (1 small box, moving both charges in the same amount) and record the electric field strength for both locations for distances between charges of $d = 1.0, 0.8, 0.6, 0.4$ and 0.2 m.

What trend do you observe? Why does this happen? What is the approximate ratio of the field strengths at the two locations (y-axis location / x-axis location) for the smallest value of d?

(3) Ring of charge. Predictions: Consider positive charge spread uniformly along a circle. Make a prediction (no detailed calculations desired here just educated guesses and maybe a brief reason why) in your notebook about the direction of the electric field at: the center of the circle; halfway from the center to the ring; just inside the ring; and outside the ring.

Arrange 24 or 28 identical positive charges in a circle of radius 1.0 m (4 large boxes in diameter). You can use the tape measure tool to help keep the charges at the correct distance, and do your best to space them evenly by eye around the circle (break it up and do a quadrant at a time, with 6 or 7 charges per quadrant.) Use a field sensor to explore the electric field more carefully. Where is the electric field a minimum? Where is it a maximum? How do the directions compare to your predicted directions? Can you formulate an explanation for why the field has the direction it does inside the ring?

(4) Line of charge. Put one negative electric charge centered vertically on the working grid. Put an electric field sensor one meter (2 large boxes) to the right. Record the field strength. Then add a pair of negative charges, one each 20 cm (2 small boxes) above and below the first charge. Record the new field strength. Repeat this process and make a table of total number of charges (1, 3, 5, 7, ...) and resulting field strength, until you have a line that has been extended to the top/bottom of the grid.

What does the resulting electric field pattern look like, particularly near the center of the line?

Move the electric field sensor closer, to 0.5 m, from the line. What is the new field strength? Move the sensor to 2.0 m from the line and record the field strength. Does the field strength vary as $1/r^2$ (like a point charge), or does your data suggest $1/r$ instead?

Plot the electric field strength at the original location (1.0 m) as a function of the number of charges in the line. (y-axis of graph = field strength, x-axis of graph = number of charges). Does it appear to be approaching a limiting or asymptotic value? If so, approximately what value? Can you predict it? (Hint: Find the linear charge density, λ = total charge/length, for your line charge, then consider what is the field due to an infinitely long line charge with the same λ, $E(r) = \frac{\lambda}{2\pi\epsilon_o r}$.)

II. Electric Field Hockey Simulation

In this simulation, (electric-hockey_en.jar) one electric charge serves as the hockey puck and is free to move. It is positive by default. Additional charges from reservoirs of positive and negative charges can be placed at fixed positions on the playing field to create an electric field that exerts a force on the puck-charge and governs its motion. These added charges are fixed in place.

The object is to guide the puck into the goal. There is a simple practice configuration, which you should play with to gain a feeling for the way the puck moves in response to the electric field. Because the electric field of each added point decreases with distance as $1/r^2$, you may find some of the behavior unexpected and harder to control than anticipated. You can turn on a vector display of the electric field pattern to the added charges and a trace of the pucks path. Leave the pucks mass at its default value of 25.

Once familiar from the practice, select difficulty level 1 (with a single barrier) and find a way to place charges to guide the puck into the goal. After a failure, try repositioning the fixed charges or adding more. You will be judged on the number of charges you need, and on the number of attempts. Print out your successful run and have your lab instructor initial it. (To print, you can copy the simulation window to the Windows clipboard with Alt-PrintScreen, open Windows Paint program (in Accessories) and paste the screen capture into Paint for printing.)

As time permits, move to difficulty level 2 and repeat. Try level 3 if you succeed at 2.

Before you leave: Make sure your lab table is clean and ready for the next group. Please don't leave random scraps of paper, calculators, water-bottles, etc., behind. Use recycling bins down the hall, not the trash, for recyclable materials.

15—Mapping fields and potentials

Object

To measure lines of constant electric potential (voltage) and use those to sketch the associated electric field lines and to develop a better understanding of the relationships connecting electric field and electric potential.

Background

A gravitational analog to electric potential can help sort out the ideas of field, potential and force. A topographic map of a surface terrain provides information in the form of lines constant elevation, h, called contour lines.

Since gravitational potential energy can be written (at least approximately) as $mgh(x, y)$ for a mass m (e.g. a ball) located at x, y, such contour lines also represent lines of constant gravitational potential energy for the ball. Following this "equipotential" line neither increases or de-

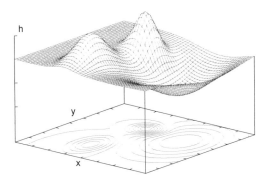

Figure 1: A surface and its topographic map showing contours of constant height.

creases potential energy. If the ball were to be released from rest from a point on a contour line, it will, of course, roll downhill. The direction it starts out will be in the direction of steepest descent from the starting point, which is *perpendicular to the equipotential contour line* and points in the direction of the force on the ball.

You could carry the ball around the landscape, release it from various locations and note the direction it begins to move. By creating a map of this initial direction throughout the terrain, "lines of force" could be drawn

111

to indicate which way a mass will tend to accelerate if released from rest throughout the landscape. By constructing lines perpendicular to these "lines of force," the equipotential (constant height) surfaces can be sketched, at least roughly.

Alternatively, if you map out the elevation of the terrain $h(x, y)$, with surveying instruments and construct the contours of constant height, you could figure the "lines of force" by drawing lines between the contours that always cross them at right angles. Having one set of lines lets you sketch the other set, at least qualitatively.

These ideas may be applied to electric fields and electric potential or voltage. The potential difference or voltage between two points is defined in terms of the electric field:

$$\Delta V = V_b - V_a = -\int_a^b \vec{E} \cdot d\vec{s}.$$

If one chooses a path in a series of small steps $d\vec{s}$ so that for each step the change in potential is exactly zero: $dV = -\vec{E} \cdot d\vec{s} = 0$, an equipotential line is traced out. To guarantee that dV is zero for each step, $d\vec{s}$ must be chosen to be exactly perpendicular to the local $\vec{E}$. This is completely analogous to walking across a terrain following a contour of constant height. Alternatively, if a positive charge q is set loose in an electric field, it experiences a force $q\vec{E}$ and accelerates in that direction - the direction of the field. Since the electric potential energy of a charge is $qV(x, y)$, the electric field pushes the charge toward lower electric potential energy, much as a ball will move to lower gravitational potential energy when released. In the electrical case, the "terrain" the charge moves over is a "electric potential or voltage surface," $V(x, y)$. While surveying instruments let you measure $h(x, y)$, you can use a voltmeter to measure $V(x, y)$.

Given a set of equipotential lines, the electric field in the space between the lines can be estimated. If the two lines represent potentials of V_2 and V_1 and they are a distance Δs apart, the magnitude of the electric field is

$$|\vec{E}| \approx \left| \frac{V_2 - V_1}{\Delta s} \right| = \left| \frac{\Delta V}{\Delta s} \right|$$

Procedure

Mapping equipotentials by hand: Lines of constant potential will be mapped for two arrangements of metal electrodes on a sheet of conductive paper. The electric potential (voltage) will be measured with a voltmeter. One of the electrodes will act as a common reference point for measuring voltages. (This reference point or level is often called "ground," just as topographic maps might report an elevation as "200 meters above sea level," as a way of providing a common and easily identified reference for height measurements.) Once the equipotentials are known, electric field lines can be sketched.

Begin by mounting a blank sheet of white paper on the plexiglas field mapping table. Make two small holes to fit the paper smoothly over the two posts. Overlay this with a sheet of carbon tracing paper, carbon side down. On top of the carbon paper put a sheet of conductive paper, which is black with grid marks. Be careful not to smudge the underlying sheet and center all layers on the posts. The disk electrodes are recessed slightly on one side. This side should be down so that the raised edge makes contact with the conductive paper. It is helpful to gently sand the raised edge to remove the insulating oxide that forms. Put one disk-shaped metal electrode on each post and clamp in place with a wing nut. Electrical connections to the electrodes are made via wires attached to the posts on the underside of the field mapping table. Connect the leads to the battery pack — black or (-) to negative (-) on the battery and red or (+) to positive (+) on the battery. The battery serves as a steady source of electric potential and electric charge. There is now a fixed potential difference (about 3 V) between the electrodes and a small electric current carries charge through the conductive paper. Since the metal electrodes are good conductors, each electrode should itself form an equipotential region, one electrode higher in voltage than the other. Points in the paper should be intermediate between the two values. We can arbitrarily assign the electrode connected to the negative battery terminal a value of $V = 0$.

Initial qualitative observations. Clip the black (-) lead of the voltmeter to the post of this $V = 0$ electrode with an alligator clip. Use the red (+) probe of the voltmeter to gently slide along a straight line starting from the $V = 0$ electrode to the other (+) electrode. Observe the voltmeter reading as you do so. Try a few other curved paths to travel between the electrodes and observe the behavior of the voltmeter.

A. Tracing equipotentials for the dipole. Return to a point where the voltmeter reports a potential of 1.50 V. (A few mV deviation is not critical in these measurements.) Press firmly (but don't puncture the paper) to mark this location through the carbon paper on the white sheet underneath. Move the voltage probe to a nearby location (1-2 cm away) that has the same voltage. Mark this point similarly. Continue tracing out points at this voltage until you have either followed a path that returns to your first point or you have reached the edge of the paper. If necessary, return to your first point and find other points nearby at the same voltage and similarly trace and mark the path of equal potential points again until you reach another edge.

Repeat this process for additional voltage readings of 0.25, 0.50, 1.00, 2.00, 2.50, and 2.75 V to generate a set of 7 equipotential contours.

Trace out the shape of the electrodes to ensure they are marked on the white sheet underneath. Disassemble the table to remove the white sheet. Mark the electrode locations on your sheet as they were connected to the battery with + or - . Each lab team member should make a copy of this data at this time.

Connect points of equal potential with *smooth* solid lines and label each contour with the corresponding voltage.

B. Two bars: a model for capacitors. Repeat this process with two long rectangular electrodes arranged parallel to one another.

Sketch in the field lines. On each equipotential map, sketch the corresponding electric field line pattern by sketching in 8 to 10 electric field lines. Begin from the 1.5 V equipotential and drawn 7 short line segments perpendicular to the contour distributed uniformly along its length. Qualitatively fill out the electric field lines by extending these line segments in a way so that they satisfy the requirement that the electric field lines are perpendicular to the equipotentials. Distinguish the field lines from the equipotentials by using a different color or using dashed lines to represent the field lines.

Mapping equipotentials with a simulation: Mapping and drawing even a few equipotentials by hand is time consuming and tedious. The *Charges and Fields* simulation from the University of Colorado PhET collection (http://phet.colorado.edu) can calculate and plot equipotential lines resulting from a charge distribution with a click of a mouse.

Figure 2: Simulation geometries modeling two parallel bars, bar and disk.

Double-click on the charges-and-fields_en.jar Java archive file in the course folder to start this simulation. It provides reservoirs of positive and negative charges ($+/-$ 1.0 nC each, n= nano = 10^{-9}), which may be dragged onto the open working area. Before doing anything with the charges: (1) Make sure the "Show E-field" box is now unchecked; (2) Check the "grid" box. The scale is such that each small box of the grid that appears is 10. cm on a side; each larger square is 0.5 m.

C. Two bars again. Construct a simulation version of the two parallel bars by arranging two lines of point charges, each containing 13 charges from the charge reservoirs. Equally space the positive charges in a vertical line toward the left side of the grid 25 cm apart on the grid and arrange matching negative charges 3.0 m away to the right of the line of positive charges.

Use the movable equipotential gadget to find a place where the potential is zero. (For the individual point charges in the simulation, the potential is given by $V(r) = \frac{1}{4\pi\epsilon_o}\frac{q}{r}$ and is zero at infinity. But with both positive and negative charges present the potential can be zero at points other than infinity.) Click the plot button on the voltage sensor and the simulation constructs the equipotential contour line of that value.

Click the "Show Numbers" box to display the voltage of the contour and repeat this process to map out the equipotential lines for $\pm 20\,\text{V}$, $\pm 40\,\text{V}$, $\pm 60\,\text{V}$, and $\pm 100\,\text{V}$. Move the potential gadget vertically between equipotentials so the numerical values on the contours don't overlap and obscure each other.

Enable the tape measure tool and measure the distance between the $\pm 20\,\text{V}$ equipotential lines near the midpoint between the lines of charges. Use this to estimate the electric field in this area via the relationship given earlier: $|\vec{E}| \approx |\frac{V_2 - V_1}{\Delta s}|$.

Turn off the tape measure or move it to an edge of the display, move the equipotential gadget to the side and print out a copy of the equipotential map. (To print the simulation, you can copy the simulation window to the Windows clipboard with Ctrl-PrintScreen, open Windows Paint program (in Accessories) and paste the screen capture into Paint for printing. For best results, in "Page Setup" choose "Orientation - Landscape and Scaling - Fit to 1 x 1 page." You may also try printing within the simulation by right clicking on the mouse and choosing print; configuring for landscape mode under printer properties again works best.)

Use an electric field sensor to measure the electric field directly within the simulation and compare your estimate from the equipotentials with this field sensor value.

Flip on the electric field display for a moment and use that to help you draw five well spaced electric field lines on your printed equipotential map connecting the two lines of charges.

D. Charged bar and disk. Move the outer 12 negative charges in your simulation together into a small cluster centered on the middle negative charge, to form a charge pattern of a (positive) bar and a (negative) disk of charge.

Map out the equipotentials (0, ±20, ±40, ±60, ±100 V) for this arrangement. Use the tape measure tool and the appropriate equipotential lines to estimate the electric field at the midpoint between the positive line of charge and the negative cluster.

Turn off the tape measure or move it to an edge of the display, move the equipotential gadget to the side and print out a copy of this equipotential map.

Use an electric field sensor to measure the electric field directly at the center between positive and negative charges within the simulation and compare your estimate from the equipotentials with this field sensor value.

Flip on the electric field display for a moment and use that to help you draw five well-spaced electric field lines on your printed equipotential map for this charge configuration.

Analysis of the hand-made potential and field maps. Return to your hand-measured equipotential maps for the two-disk (dipole) configuration and parallel bar electrodes.

(1) Estimate the electric field magnitude $|\vec{E}|$ near the $V = 1.5\,\text{V}$ equipotential near the midpoint between the disk electrodes, based on the distance to nearby equipotentials.

(2) Measure the distance between the two disks. Assuming the disks can be modeled as point charges of opposite sign, $\pm q$. What value of q would produce the net electric field determined in the previous question? That is to say, estimate the amount of charge on each disk and compare it to the amount of charge used in the simulations.

(3) Estimate the electric field strength in the middle of the map between the two parallel bar electrodes.

(4) If electric field lines and equipotential contours are supposed to cross at right angles, what general trends can you say about how electric field lines are oriented at the surface of a very good electrical conductor, such as the metal electrodes? How are equipotential surfaces arranged near the surface of a metal?

Before you leave: Make sure your lab table is clean and ready for the next group. Please don't leave random scraps of paper, calculators, water-bottles, etc, behind. Use recycling bins down the hall, not the trash, for recyclable materials.

16—Simple electric circuits

Object

To build and observe the operation of simple electric circuits; to learn measurement methods for electric current and voltage using ammeters and voltmeters.

Background

Electric charges move through electrical conductors in response to electric fields or, equivalently, differences in electric potential or voltage. The electric current is the rate charge passes a point in the circuit, $I = \frac{\Delta q}{\Delta t}$. Electric current is measured in coulombs per second, or amperes (A). One coulomb passing the measuring point in one second is a current of one ampere. Often in laboratory settings electric currents are more conveniently expressed in milliamperes, 1 mA = .001 A.

The electric circuits in this lab are built with batteries, light bulbs, switches and wires. The batteries are a source of electric potential energy and produce a (nearly) constant voltage difference. The characteristics and arrangement of the light bulbs will determine how much current flows. Meters for measuring current (ammeters) and voltage (voltmeters) will be added after preliminary observations.

A set of circuits will be built with real components and with a simulation that allows the use of ideal components. Both qualitative observations and quantitative measurements will be made of the electric currents and voltage differences. The circuits here are often described as direct current (DC) circuits, since the current that flows through each circuit moves in one direction through each component at all times. Circuits using time-dependent currents, alternating current or AC circuits, will be encountered in later labs.

The circuits are presented in electrical diagrams or schematics that use standard symbols for various components. The components used here are shown below. The wire example also shows a *junction*, where electrically

117

connected wires are indicated by the solid circle. Such a point is a fork in the road for a charge moving through the circuit.

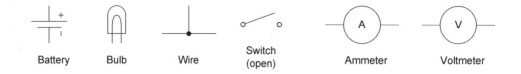

Figure 1: Symbols for circuit components used in this lab.

In building circuits, it is helpful to arrange the actual components as closely as possible in the same layout as in the schematic. This allows the easiest detection of mistakes in wiring. Keep your circuit area clear of unused wires, components and meters. You should trace out each circuit you build with a finger to make sure what you build corresponds to the electrical schematic.

Procedure

Qualitative observations. Using the real components at the table, but without using any meters yet, begin by building the simplest of our circuits using the battery pack, one light bulb and the switch, along with necessary wires to make the connections, shown in Fig. 2. Place your components as shown in the schematic as closely as possible. The switch is "normally open." When the switch is pushed to bring the two sides into contact, it is said to be *closed*. What happens in your circuit when the switch is closed?

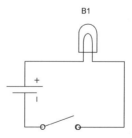

Figure 2: Simple initial circuit, C1.

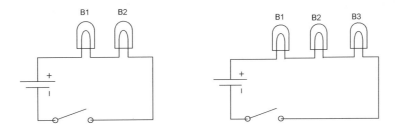

Figure 3: Series circuits S2 and S3.

Continue by building the two *series* circuits in Fig. 3. For each circuit describe the circuit performance and rank the brightness of the bulbs.

Return to the starting circuit and add a second bulb in *parallel* with the original to form circuit P2, as shown in Fig. 4. Rank the relative brightness of the bulbs in when the switch is closed. Then add a third bulb in parallel to form circuit P3. Rank the brightness of the three bulbs.

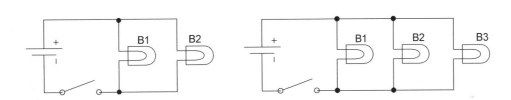

Figure 4: Parallel circuits P2 and P3.

Virtual circuits - simulations. Next, use the computer to build comparable virtual circuits using the Circuit Construction Kit from the PhET collection of the University of Colorado (http://phet.colorado.edu). In the course folder open the "circuit construction kit" java (.jar) file. This provides batteries, wires, switches, and light bulbs that may be dragged into the work area and connected together to form a circuit. Wires can be stretched to a desired length and will automatically make a connection when an end is brought near another connection point. Components can be rotated into more convenient orientations after being dragged to the working area by grabbing a connecting terminal with the mouse. Once a circuit is wired up, right-clicking on a component will present options to delete the object. Right-click on a connection point between two components and you can choose to break the connection to modify the circuit.

Rebuild the simple circuit in figure 1. Grab the switch handle and close the switch completing the circuit and watch the simulation. The bulb lights and the simulation shows the motion of the charges in the circuit that constitutes the electric current.

Important note: The simulation accurately reflects the direction negatively charged electrons move in the circuit: out of the negative terminal of the battery and into the positive terminal. In most circuit analysis we regard electric current as the motion of positive charges flowing in the opposite direction; a negative charge flowing right to left in a wire is the same electric current as a positive charge flowing left to right. Most textbooks treat electric current by convention as the flow of positive charges from the positive (high potential) battery terminal and into the negative (low potential) terminal. Please keep in mind that the simulation shows what electrons are actually doing.

Measuring electric current. Electric current is measured with an ammeter. In order for the meter to measure the current, the electric charges must flow through the ammeter. Insert an ammeter into the circuit to measure quantitatively the current in the simulation. (Click the ammeter box under "Tools" and drag an ammeter into the work area. Then break existing connections to wire the ammeter into your circuit.

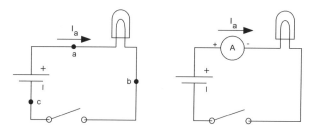

Figure 5: Use of ammeter to measure current passing point (a). The arrow illustrates the direction of motion of conventional positive charges.

What current is flowing in the circuit?

Leaving the ammeter at point a, add a second bulb in parallel with the first to form a virtual version of circuit P2 as shown in Fig. 6. What is the current reported by the ammeter now?

Look carefully at the motion of the simulated charges. Compare the speed of the charges as they enter the ammeter with the speed of the charges moving through each bulb. Watch closely at the junction where the charges from the bulbs B2 and B1 meet up and continue on to the ammeter. What do you think the current through bulb B1 is? What about the current through B2?

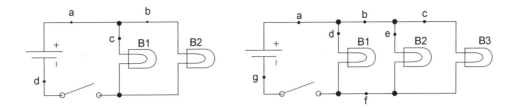

Figure 6: Parallel circuits P2 and P3 and possible current measurement points.

Check one of these by adding a second ammeter to your circuit at point (b) or (c) to measure one of these currents.

Add a third bulb, B3, in parallel with the first two to make a virtual version of circuit P3. What effect, if any, does connecting B3 have on the speed of the charges moving through B1 and B2? What is the effect on the speed of the charges at the battery? What is the current reported by the ammeter at (a) now? Again look closely at how the moving charges divide and combine at various junctions of three wire segments.

Summarize these observations about how electric currents behave at junctions. How does the total current arriving at a junction compare to the total current leaving a junction of three wire segments?

How much charge does one of the the moving balls carry? Each of the moving balls in the simulation represents many, many electrons. Watch closely at one ammeter terminal. Record the ammeter reading and count the number that enter the ammeter during some time interval (10 seconds might be reasonable). From this, determine how much charge (in coulombs) a single ball represents. Show your reasoning. Recall that 1 ampere = 1 coulomb per second.

Bulbs in series Build a virtual version of circuit S2 (in Fig. 7) of two bulbs in *series*. Insert an ammeter at point (a). How does the current compare to the current when just a single bulb was present earlier? Insert an ammeter at point (c). How does the current at a compare to the current at (c)? What then should be the current at (b)?

Leave the ammeters in place and add a third bulb after the 2nd ammeter to form circuit S3. What is the new current at point (a)? And at (c)?

Summarize these observations:

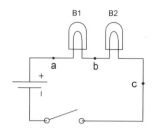
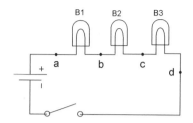

Figure 7: Series circuits S2 and S3; possible current measurement points.

When several light bulbs are arranged in series, what can you say about the current flowing through each bulb?
How does the current in the circuit change as the number of bulbs in series increases?

Measuring currents with a real ammeter

REAL AMMETERS ARE EASILY DAMAGED OR DESTROYED BY ALLOWING CURRENTS TO FLOW THAT ARE LARGER THAN THE FULL-SCALE VALUE FOR A GIVEN SETTING. THEY ARE THE LEAST DURABLE OF ELECTRICAL INSTRUMENTS YOU WILL USE IN LAB. **When using ammeters always begin measuring using the largest current range available** (E.G., 5 A HERE). IF THE METER PRODUCES A SMALL READING OR DEFLECTION, THEN CHANGE THE RANGE TO THE NEXT MORE SENSITIVE (SMALLER RANGE) SETTING, E.G., 0.5 A. DO NOT USE A RANGE THAT WILL SEND THE METER READING BEYOND THE FULL-SCALE VALUE. ALWAYS DOUBLE CHECK YOUR WIRING AND RANGE SELECTION BEFORE OPERATING YOUR CIRCUITS.

Return to the real components on the table. Build circuit S2 (Fig. 7). In our real circuits and real ammeters, we think in terms of conventional current: the flow of positive charge. Insert an ammeter at point (a), taking care to have the positive terminal (often the red connector) connects to the positive terminal of the battery so + charges from the battery flow into the red terminal, through the ammeter and out the negative (black or 'common') ammeter terminal. (Remember, conventional current flows from the positive terminal of the battery, through the circuit, to the negative terminal of the battery.) Measure and record the current at (a), then insert the ammeter in turn at (b) and (c) to record the currents at those points. The values will be different than found in the simulation (whose ideal batteries and bulbs aren't available in any store), but they should reflect the same relationship: how do currents at (a), (b), and (c) compare in the real circuit?

Again with the real components, build circuit P2 (Fig. 6). Insert the ammeter successively at points (a), (b), and (c) to measure those currents. For circuit P2, based on the measurements and given their limited precision, what is the likely mathematical relationship between I_a, I_b, and I_c.

Voltage measurements. Measuring potential difference or voltage is done with a voltmeter. Unlike ammeters measuring current at a single point in the circuit, voltmeters compare the electric potential at *two different points* in the circuit. The voltmeter leads or probes can be attached or moved without disassembling the circuit, as was necessary with the ammeter. The voltmeter reports the voltage difference, ΔV, between the two probe locations, $V_+ - V_-$ or $V_{red} - V_{black}$. This difference is usually called simply "the voltage" or "the voltage drop" between the points. It is a good idea to use color-coded probes or leads with the voltmeter to keep the proper sense of polarity in the measurements. Let's represent these differences by a simple notation: $V_{AB} = V_A - V_B$, so that V_{AB} would be measured by placing the red (+, or high) probe at point A and the black (- or low) probe at point B. The real voltmeter used here is a digital multimeter that can be used to measure voltage, current, and resistance. Be sure to select the appropriate function - *DC voltage* measurement, not AC - and a suitable range for the measurements at hand. Ask your lab instructor for assistance in interpreting the various icons on the meter if necessary. Be sure *not* to use a current-measuring scale!

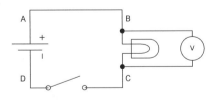

Figure 8: Measuring the voltage drop V_{BC} across a light bulb.

Begin with the simple one-bulb circuit again (in the real world, not the virtual circuit) and measure the voltage drop V_{BC} directly across the light bulb as in Fig. 8 with the switch closed. Also measure the voltage drops along the wire from the + battery terminal to the light bulb, V_{AB} and from the bulb back to the − terminal of the battery, V_{CD}. Measure V_{DA}, and V_{AD} (reversing the roles of red and black leads). Be sure to measure all of these with the switch closed and the circuit functioning. Re-measure the voltages when the switch is open. List all measurements (switch closed and open) in a table. Calculate the sum of the voltage drops found moving the voltage probes once completely around the circuit:

$$\Sigma V = V_{AB} + V_{BC} + V_{CD} + V_{DA} = ?$$

Or equivalently, how does $V_{AB} + V_{BC} + V_{CD}$ compare to V_{AD}? Does it appear reasonable to neglect the voltage drops V_{AB} and V_{CD} along the wire segments compared to voltage drops across the bulb?

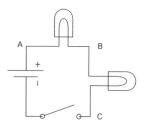

Figure 9: Voltage reference points for series circuit S2.

For the two series circuits S2 and S3 measure the voltage drops across each light bulb and the battery again. You need not measure the separate small voltage drops along wire segments this time, just pick useful points as in Fig. 9 to produce the complete set of voltages, V_{AB}, V_{BC}, and V_{CA} around the circuit. Make all measurements with the switch closed, of course. In each circuit compute the sum of the voltage drops around the circuit again.

Measure the voltage drops across each bulb for the parallel-bulb circuit, P2 as well as the battery voltage (with switch closed!). How do the voltage drops across the bulbs compare to each other and to the battery voltage?

Mixed series and parallel circuit. Consider circuit S2 in operation. Make a table predicting what you think will happen to B1 and B2 when a third bulb B3 is added in parallel to B2, as shown in Fig. 10. Do this by listing in your notebook whether B1 and B2 will become brighter, dimmer, or stay the same when B3 is added in parallel with B2. Also predict the relative brightness of the three bulbs after B3 is added.

After writing your predictions, connect B3 in parallel with B2 while the switch is closed and report what actually happens to the brightness of each bulb in the circuit.

Measure the voltage drop across each bulb. How has each value changed when B3 is added compared to the simple series circuit of B1 an B2 (for

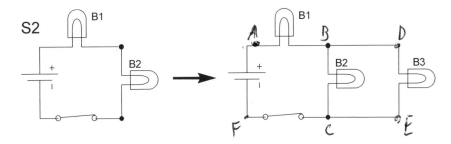

Figure 10: Converting S2 to a combined series and parallel configuration by adding B3.

which current and voltage measurements are already made earlier.) Has the current through B1 increased or decreased with the addition of B3? By how much?

Current flowing through the voltmeter. For the simple one-bulb circuit, re-build Fig. 5 with the ammeter. Observe the current flowing when the switch is closed. Attach the voltmeter as in Fig. 9 directly across the lightbulb only, and again measure the current. Does adding the voltmeter significantly increase the current provided by the battery?

Use the voltmeter to measure the voltage drop across the ammeter. How does it compare to the voltage drop across the bulb?

Mistakes with ammeters. In the simulation (not with the real world ammeter!) wire up an ammeter *in parallel* with the light bulb in the simple one bulb circuit and close the switch. What happens? This is an an example of a "short circuit," where the + terminal of the battery has been connected almost directly (as if with a simple piece of wire) to the − terminal. The consequences with a real ammeter may range from a destroyed meter (now an expensive paperweight) to the need to replace a fuse in the device (if you're lucky). This highlights the importance of understanding that voltmeters and ammeters measure different quantities and must be wired into a circuit differently.

Questions. Generalize and summarize your observations on circuits by answering the following questions.

1. What can you say about the electric current that flows through two or more bulbs connected one after another in a series?

2. What can you say about the currents flowing into a junction of three wires compared to the currents that flow out from the junction?

3. What can you say about the voltage drops across light bulbs in parallel?

4. If you were to follow a single (positive) electric charge as it flows around any complete path in the circuit, starting from the positive battery terminal returning to the negative battery terminal and added up each voltage drop the charge experienced, how would the total voltage drop from the light bulbs compare to the voltage supplied by the battery?

5. How does the brightness of the bulb depend on the current flowing?

6. If a light bulb is not glowing, does that mean there must necessarily be no current flowing?

Before you leave: Make sure your lab table is clean and ready for the next group. Leave equipment neatly organized for the next lab. Please don't leave random scraps of paper, calculators, water-bottles, etc., behind. Use recycling bins down the hall, not the trash, for recyclable materials.

17—Resistance and Ohm's law

Object

To test Ohm's law with a carbon resistor, measure resistances in series and parallel, and to measure the current-voltage characteristics of a light bulb.

Background

For many conductors, especially metals, the current flowing through a device is proportional to the voltage difference applied to the device: the ratio of the voltage to the current is a constant. This is Ohm's law. This ratio for a particular device *defines* its resistance: $R = V/I$. For many materials under ordinary lab conditions, the ratio *is* practically constant, and the device is then described as ohmic. In some situations the ratio V/I of a device may vary with changes in the conditions of the measurement such as large changes in the applied voltage or temperature. These properties turn out to be useful in their own right.

In this experiment you will test whether a carbon resistor (a common electronic component) obeys Ohm's law by measuring the current and voltage and calculating the ratio of V/I. If Ohm's law holds, the ratio should be a constant.

To measure the resistance of a device one needs simultaneous measurements of the current passing through the device and voltage drop across the device. Simultaneous measurements can be tricky: when a voltmeter or ammeter is inserted in a circuit its presence will affect, to a greater or lesser extent, the currents and voltages throughout the rest of the circuit. The presence of one meter will affect the reading of the other—preferably only slightly.

There are two basic circuit configurations for the voltmeter and ammeter connections to measure the resistance of a device. These are shown in Fig. 1. In configuration (a) the ammeter measures the true value of the current in the unknown resistor, but the voltmeter measures the voltage drop across the series combination of the resistor and the ammeter, not the voltage across the

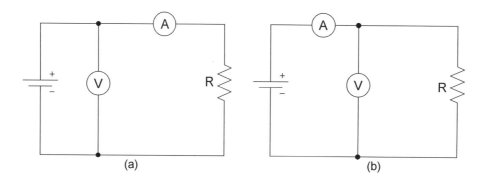

Figure 1: Circuits for measuring resistance

resistor alone. If the resistance of the ammeter is known, the voltage drop across the ammeter can be calculated and the voltmeter reading corrected to give a better value of the voltage drop across the resistor itself.

In configuration (b) the voltmeter correctly measures the voltage across the resistor, but now the ammeter reports the sum of the currents through the resistor *and* the voltmeter. If the voltmeter's internal resistance is known, the current diverted by the voltmeter can be calculated and the ammeter reading corrected. Unless the resistor has a very high value, the effect of the typical digital voltmeter is negligible.

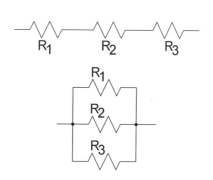

When two or more resistors are connected together, they can be replaced by a single functionally *equivalent* resistor. Combinations of resistors can be broken down into two kinds: series and parallel. For two or more resistors connected in series, the equivalent resistance is the sum of the individual resistances:

$$R_{eq}^{(series)} = R_1 + R_2 + R_3 + \cdots.$$

Figure 2: Series and parallel resistors.

For two or more resistors in parallel, the equivalent resistance of the combination is given by

$$\frac{1}{R_{eq}^{(par)}} = \frac{1}{R_1} + \frac{1}{R_2} + \frac{1}{R_3} + \cdots.$$

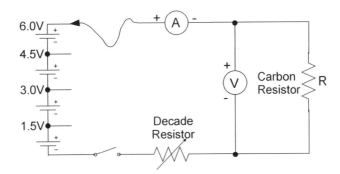

Figure 3: Measuring R of a resistor made of a carbon film.

Procedure

Resistance measurements. The resistors provided are made from carbon-based composite material. Identify the values of the resistors provided by interpreting the color coded bands. (See the appendix on the resistor color code to decipher the values.) Record these as the nominal resistances.

Connect the circuit as shown in Fig. 3 using an 18 kΩ resistor. The ammeter has a 200 microampere maximum range. Use the digital multimeter set to measure DC voltages (VDC). The voltages are supplied by a pack of four batteries connected in series. The decade box is an adjustable resistance, which is present to limit the current in the circuit and protect the ammeter. Be sure to set it initially for 15 kΩ. Have the instructor check your circuit before closing the tap key switch.

CAUTION: TO PROTECT THE AMMETER, DO NOT DRAW CURRENTS GREATER THAN 200 µA FROM THE BATTERY PACK. MAKE SURE THE TOTAL NOMINAL RESISTANCE IN THE CIRCUIT IS MORE THAN 30,000 Ω BEFORE CLOSING THE SWITCH FOR THE FIRST TIME. BE SURE THE + TERMINAL OF THE AMMETER CONNECTS TO THE + TERMINALS OF THE BATTERY PACK.

To test Ohm's law, connect the 6 volt terminal of the battery to the ammeter lead. Adjust the decade resistance box to give a current reading of 200 µA (i.e. full-scale).

Record the current and voltage readings for connections to the 1.5, 3.0, 4.5, and 6 volt terminals of the battery. Also record I and V when the switch is open. Include this point in your graphs.

Plot I as a function of V for this resistor—that is put I along the y-axis and V along the x-axis. (This choice of axes is the standard way of presenting a device's electrical characteristics, even if backward from what might be the simpler reverse choice here.) Is the graph linear? If so, then Ohm's law applies and $I = \frac{V}{R}$. Find the slope of the line, and from the slope find the resistance. For each measurement calculate V/I. Do these values agree with the resistance found from the slope and with the nominal value from the color code, within the 5% tolerance indicated by the color code?

Replace the 18 kΩ resistor with a 12 kΩ resistor (identified by its colored bands). Do not change the decade box. Using the 4.5 V battery terminal (or 3 V, if necessary to keep the current on scale), make a *single* (V, I) measurement and calculate the actual resistance. Make a similar measurement of a 47 kΩ resistor.

Series resistors. Put two 18 kΩ resistors in series as R in the circuit. (Do not adjust the decade resistor box.) Using the 6 V battery terminal, what are the new voltage and current readings for the 2-resistor series combination? Find the equivalent resistance from the ratio $R_{eq} = V/I$.

Replace the second 18 kΩ resistor with the 12 kΩ resistor already measured, and again find the equivalent resistance of the pair using the 6 V terminal of the battery pack. Also use the voltmeter to measure and record the voltage drop across each resistor individually. Use these individual voltages and the nominal (color-code) resistance to double check the current flowing through each resistor using Ohm's law.

Again replace the second resistor, this time with a value larger than the original resistor (e.g. 33 kΩ or 47 kΩ according to the color code), and again find the total resistance of the pair.

Do your total or equivalent resistances agree with the rule for series resistors? Tabulate the measured and predicted values

Parallel resistors. With the ammeter lead connected to the 1.5 V battery terminal, add a second equal resistor in parallel to the original one 18 kΩ resistor circuit. What are the new current and voltage readings? Use these measurements to calculate the equivalent resistance of the parallel combination.

Make measurements with a third identical resistor in parallel with the first two.

Measure the equivalent resistance of two resistors in parallel using 18kΩ and 12kΩ resistors.

Measure the equivalent resistance of two resistors in parallel using 18kΩ and 33kΩ resistors.

Does the combined equivalent resistance of parallel resistors found in your trials agree with the value predicted from the rule for parallel resistors? List both measured and predicted values.

Will the equivalent resistance of two or more resistors in parallel always be less than any of the individual resistors?

Effects of meters. Return to the original circuit of Fig. 3 with an 18kΩ resistor. To study the effect of the meters on the measurements, with the battery connection at the 6 volt terminal and a full scale ($200\,\mu$A) meter reading, move the voltmeter's "+" lead from the "−" to the "+" terminal of the ammeter, corresponding to configuration (a) in Fig. 1. There should very little or no detectable change in the ammeter reading but a measurable change in the voltmeter reading.

Modern digital voltmeters usually have very high internal resistance ($> 10^7\,\Omega$). The current diverted by the voltmeter (as in Fig. 1(b)) then is expected to be less than a μA at 6 volts. The difference in the voltmeter readings is due to the voltage drop across the internal resistance of the ammeter. From this voltage change observed between configurations (a) and (b) and the current, calculate the internal resistance of the ammeter.

Non-linear behavior. Measure the $I(V)$ curves for light bulbs using a modified version of the original circuit configuration of Fig. 3. Make the following changes to produce and measure the higher currents of this part:

Replace the battery pack with the battery eliminator/power supply, which can provide larger currents than the battery pack and gets its power from the wall outlet. Set the voltage from the battery eliminator to its *lowest* setting.

Replace the low-current ammeter (200 μA maximum) with a more robust meter capable of handling at least 500 mA. Replace the carbon resistor with two small light bulbs connected in series.

Set the decade resistor box to zero, or better, remove it from the circuit entirely (less clutter!). The bulbs draw approximately 250-300 mA at 3 V; by using two bulbs in series, the total resistance is increased and more voltage settings of the power supply can be safely used without burning out the light bulbs.)

Be sure to keep the switch in your circuit.

With the battery eliminator set at its lowest setting, record the current flowing through the series combination and the voltage drop across each bulb in the series individually. **Caution: When changing the voltage setting on the battery eliminator/power supply, be sure the switch in your circuit is open and no current is flowing,** $I = 0$. Stepping through each available voltage setting up to a maximum of 5 V on the battery eliminator, repeat the measurements of V and I for the bulbs.

Compile a table of your V, I measurements for each bulb, and for each pair compute R for each bulb. Recall that you also have data for one additional point you probably haven't explicitly recorded: $I = 0$ when $V = 0$. (If in doubt, check what the meters read when the switch is open!) Include this point in your table and graphs! Make a graph with plots of $I(V)$ for each bulb.

How does the resistance of a light bulb change with temperature? How does this show up in the graph?

Digital multimeter as ohmmeter. In general you cannot use the multimeter reliably as a stand-alone ohmmeter with components still connected in a circuit. Take one of the light bulbs out of the circuit completely. Switch the multimeter to measure resistance (Ω). In this mode the meter sends a small current through the light bulb and measures the voltage needed to do this, all behind the scene, and reports the resistance. What is the resistance of your light bulb according to the multimeter in resistance mode?

Try this with your original 18kΩ resistor, too, and record its value.

Before you leave: Make sure your lab table is clean and ready for the next group. Leave equipment neatly organized for the next lab. Please don't leave random scraps of paper, calculators, water-bottles, etc., behind. Use recycling bins down the hall, not the trash, for recyclable materials.

18—Capacitors

Object

To study the behavior of capacitors in circuits by observing the decay in voltage across a capacitor in an RC circuit; finding the time constant of the RC circuit and the capacitance from the measured time constant; measure the current and voltage during charging through a series resistor; measure voltage and charge on capacitors in series and parallel combinations.

Background

A capacitor stores electric charge. A simple configuration for a capacitor is two parallel metal plates. The amount of charge stored is proportional to the voltage difference V between the plates. The charge stored on one plate is $Q = CV$. An equal amount of opposite charge is induced on the other plate. The capacitance C depends on the area and separation of the plates and any material (dielectric) between the plates.

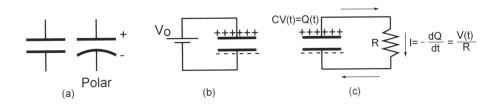

Figure 1: (a) Symbols for capacitors; (b) charging a capacitor with a battery; (c) discharging a capacitor through a resistor. Polar or electrolytic capacitors must always be connected in circuits so the + terminal is at the higher voltage.

Fig. 1(a) shows typical symbols used to represent capacitors in electrical schematics. Attaching a capacitor to a battery stores charge on the capacitor plates, as in Fig. 1(b). The stored charge can be drained by connecting a resistor to the capacitor as in Fig. 1(c). The rate at which charge flows off the capacitor plates is equal to the current through the resistor. The negative sign in $I = -\frac{dQ}{dt}$ follows from the observation that a *decrease* in the stored charge Q on the positive plate represents a *positive* current through the resistor in the direction indicated by the arrow.

When a capacitor is discharged through a resistor R the voltage across the resistor is $V_R = IR$. The same voltage appears across the capacitor, $V_C = Q/C$. The capacitor acts like a poor battery that "dies" quickly by giving up all its stored charge. The voltage as a function of time can be found after finding the charge as a function of time:

$$
\begin{aligned}
V_R(t) &= V_C(t) \\
I(t)R &= \frac{Q(t)}{C} \\
-R\frac{dQ}{dt} &= \frac{Q}{C} \\
\frac{dQ}{dt} &= -\frac{1}{RC}Q.
\end{aligned}
$$

This equation can be solved easily by re-arranging and integrating:

$$
\begin{aligned}
\frac{dQ}{Q} &= -\frac{1}{RC}dt \\
\ln Q &= -\frac{t}{RC} + B \\
Q(t) &= e^{-\frac{t}{RC}+B}.
\end{aligned}
$$

The constant B is the arbitrary constant of integration. If at $t = 0$ the capacitor has a charge Q_o, then $Q_o = e^B$. B is fixed by the initial charge on the plates. The charge decays exponentially:

$$
Q(t) = Q_o e^{-\frac{t}{RC}}.
$$

Since $V_C(t) = Q(t)/C$, the voltage also decays following the same form, with $V_o = Q_o/C$:

$$
V_C(t) = V_o e^{-\frac{t}{RC}}, \tag{1}
$$

A typical exponential decay is shown in Fig. 2(a). The product RC is called the time constant of the circuit and is often denoted as τ. When an amount of time equal to one time constant has elapsed from the start of the discharge ($t = \tau = RC$), the voltage will have dropped to $V_o e^{-1} \approx 0.37\,V_o$. After two time constants have elapsed, $t = 2\tau$, the voltage has fallen to $V_o e^{-2} \approx 0.14 V_o$. When the $\ln V_C$ is plotted against time (Fig. 2(b)), a straight line graph is expected with the slope of the line equal to $-1/RC$. To see that the slope of the $\ln V_C$ vs. time graph is $-1/RC$, take the natural logarithm of $V_C(t)$ in Eq. 1:

$$
\ln V_C = -\frac{1}{RC}t + \ln V_o. \tag{2}
$$

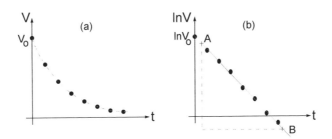

Figure 2: Voltage decay of discharging capacitor.

This is in the standard form of a linear equation, $Y = mX + b$, if we identify $Y = \ln V_C$, $X = t$, the slope $m = -1/RC$, and the Y-intercept $b = \ln V_o$

The product RC can be obtained in several ways. A "quick-and-dirty" method for a rough estimate is to make a plot like Fig. 2(a) and draw a smooth curve through the data. Calculate $V_o e^{-1} = 0.3679\, V_o$ and read off from the plot the time t when V_C reaches this value. Then $RC = t$.

A second and more careful graphical method uses a graph like Fig. 2(b). This graph provides a more conclusive test of the exponential nature of the discharge by producing a straight line if Eq. 2 is correct. From (t, V) data $(t, \ln V)$ is tabulated and plotted. A straight line is drawn through the data that best represents the experimental data. Select two points (**not** data points) near the ends of the straight line of Fig. 2(b). Call these points A and B. Read from the axes the values of $\ln V$ and t that correspond to these points and calculate the slope from the expression

$$\text{slope} = \frac{-1}{RC} = \frac{\ln V_B - \ln V_A}{t_B - t_A}. \tag{3}$$

Alternatively, you can find the best slope from a least squares (linear regression) fit of a straight line to the $\ln V$ vs t data.

Once the time constant $\tau = RC = -\frac{1}{\text{slope}}$ is found and if R is known, C can be determined:

$$C = \frac{\tau}{R} = -\frac{1}{R \cdot \text{slope}}.$$

Procedure

Exponential discharge in an RC circuit. Wire up the circuit shown in Fig. 3. Use the capacitor with the larger capacitance labeled on it. Since capacitor tolerances are often very loose, this value is more likely to be a rough estimate of its true capacitance ($\pm 20\%$ or worse). Use the computer to measure the voltage drop across the capacitor.[1]

Start with $R = 1000\,\Omega$ set on the decade resistance box. It is important that the capacitor is connected with its + terminal to the + side of the battery/voltage supply. Connecting them with reversed polarity can damage or destroy the capacitors. If in doubt, have the lab instructor check your circuit before closing the switch to charge the capacitor.

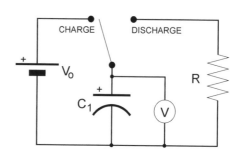

Figure 3: Monitoring discharge of a capacitor through a resistor.

Before connecting the voltage measuring leads from the computer interface to the circuit, short these leads together—connect the red and black wires directly to each other. The voltage readout in *Logger Pro* should be nearly zero. If the voltage readout is not 0.000 ± 0.006 V, use the "Zero" function in *Logger Pro* to cancel this small offset. Then connect the voltage leads to the capacitor as indicated in the circuit diagram.

Put the double-throw switch in the charging position and wait for the voltage reading to give a steady reading for a minimum of 15 seconds. Record this initial voltage.

Now start data collection and flip the switch to the discharge position Allow the measurements to continue until the capacitor voltage has dropped below 0.10 V, or for 200 seconds, which ever comes first.

<u>Estimate of the time constant</u>. Make a quick estimate of the decay time constant τ by the "quick and dirty" method descibed earlier: examine the graph of $V_c(t)$ and find out how long it takes for V_C to drop from its initial value to about $0.37V_o$.

<u>Careful determination of the time constant</u>. If a plot of $\ln V_C$ as a function of t produces a straight line, you can conclude the discharge follows the exponential described by Eqs. 1 and 2. Carry out a linear fit to this data.

---

[1]Connect the voltage measuring leads to channel 1 of the LabPro interface. Use the *Logger Pro* file "RC-discharge.cmbl" for collecting data. Live readouts of the voltage appear near the bottom of the *Logger Pro* window.

Choose the data you fit carefully—there will likely be small deviations from linear behavior at low voltages ($\sim 0.30\,\text{V}$ and below) that are not immediately apparent on the screen.[2] Record the slope and intercept of your fit, report the time constant τ. Measure the value of the discharge resistor using the resistance scale of the digital multimeter—you must temporarily disconnect the decade box from your circuit to measure it with the DMM. Use this value of R and your slope to find C.

Store the data (Ctrl-L) so it can be compared with the next measurements.

Set $R = 500\,\Omega$. How do you expect this to change the plots when you repeat the measurements? Carry out the discharge measurement again—charge the capacitor until the voltage is stable for at least 15 seconds, start data collection and simultaneously flip the switch to the discharge position. Repeat the determination of C. Store the data.

Repeat with $R = 200\,\Omega$. Put all three data sets on a single plot and include this plot in your notebook.

Find the average or your three capacitance measurements. Based on the spread in values, what do you estimate the uncertainty in your value of C_1 to be?

Second capacitor. Repeat this procedure with the capacitor of lower nominal value, but use only $R = 1000\,\Omega$. Include a graph of this data in your notebook. Find C_2.

Charging the capacitor. You will measure the current flowing to the capacitor and capacitor voltage as functions of time as the capacitor is charging up. Build the circuit shown in Fig. 4. Use the larger capacitor. The decade resistance box is now in series with the battery and an ammeter.[3] The discharge side of the double-throw switch is now simply a short circuit to allow you to rapidly discharge the capacitor completely.

Set $R = 20\,\Omega$. Discharge the capacitor fully – for at least 30 seconds. With the capacitor fully discharged and still shorted, so that both $V_C = 0$ and $I_C = 0$, zero both sensors in *Logger Pro*.

---

[2]While *Logger Pro* has the ability to fit the (t, V) data directly to the form of Eq. 1, it it much harder to determine which portion of the data accurately follows the exponential decay when looking at the $V(t)$ plot. The $\ln V(t)$ plot demonstrates which points *are* exponential, not merely close to exponential and provides a superior basis for selecting data to use to find the time constant and C.

[3]The ammeter is a current probe sensor that should be connected to channel 2 of the LabPro interface. Pay attention to the intended direction of current flow through the sensor. The red terminal of the current probe should connect to the positive terminal of the battery. The file "RC-charge.cmbl" is configured to collect data from the voltage and current probes.

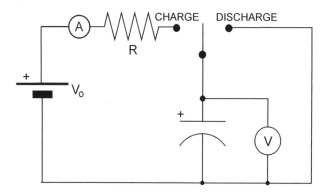

Figure 4: RC charging measurements.

Before collecting charging data *predict* the initial current that you expect to flow once the switch is thrown to the charging position. Show a supporting calculation of your predicted I_o. What do you expect the $V_C(t)$ and $I_C(t)$ curves to look like? Sketch them qualitatively in your notebook before measuring.

Start data collection in *Logger Pro* and after a couple seconds, flip the switch to the charging position. (This *Logger Pro* file uses triggering capabilities so that data are not stored until the capacitor actually begins to charge and V_C rises above 10 mV. You need not flip the switch simultaneously with starting data collection.) Record the charging process for 50 s.

Is the initial current near your predicted value? Do the plots look like your predictions?

Store the data (Ctrl-L) and repeat the measurements for $R = 50\,\Omega$ and $R = 100\,\Omega$.

Find the area under the $I_C(t)$ curve for the $R = 20\,\Omega$ run with *Logger Pro*. What does this integral represent? Compare this to CV_C, with V_C being the final voltage at the end of the charging process.

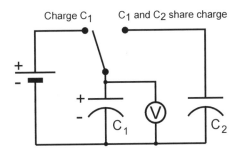

Charge C_1 C_1 and C_2 share charge

Figure 5: Parallel capacitor circuit.

Parallel and series capacitors. Rewire the circuit to the configuration shown in Fig. 5, so that when connected by the switch, the capacitors will be in parallel. Be sure that C_2 is fully discharged *before* it is placed in the circuit: connect the two terminals of C_2 directly together with a wire ("short circuit" the terminals together) for 10 to 20 seconds; then remove the shorting wire and put C_2 in your circuit.

Throw the switch into the charging position and wait to get a steady voltage reading—use the digital multimeter in DC volts mode—across the capacitor C_1. Record this voltage (V_o). Throw the switch into the charge sharing position. The two capacitors are now in parallel. The voltmeter reads the voltage across the parallel combination. The original charge on C_1 is now shared by the two capacitors with no loss. Record the new voltage V across the parallel combination.

The charges on the two capacitors are given by $Q_1 = C_1 V$ and $Q_2 = C_2 V$ and the total charge is given by $Q_1 + Q_2 = C_1 V_o$, the charge originally stored on C_1. Solve these equations for V across the parallel combination in terms of C_1, C_2, and V_o:

$$V = ? \, V_o$$

Calculate the expected value of V using the values of C_1 and C_2 you have determined earlier. Do not use the values labelled on the capacitors. Compare this expected voltage to your measured value by calculating the % difference.

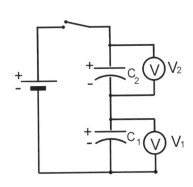

Figure 6: Series capacitor circuit.

In a *series* combination the charge stored must be the same for each capacitor and the voltages across capacitors are inversely proportional to the capacitance. Note that the two "intermediate" capacitor plates where C_1 and C_2 are connected together can only exchange charge with one another, so any charge one has gained, the other must have given up: $Q_1 = Q_2$. In terms of C_1 and C_2, what is the ratio expected between V_1/V_2?

$$\frac{V_1}{V_2} = ?$$

Predict the ratio of V_1/V_2 from your values of capacitances C_1 and C_2.

Build the circuit shown in Fig. 6, where the capacitors are connected in series. Again be sure that each capacitor is completely discharged before wiring the circuit. Close the switch to charge the series combination of the two capacitors.

Using the digital meter measure the voltage drops across each of the two capacitors. Compare the ratio $\frac{V_1}{V_2}$ to the predicted value by computing the % difference.

Additional analysis. When the capacitor is charged through the resistor, the voltage across the capacitor is given by $V_C(t) = V_o(1 - e^{-t/\tau})$ and the current is given by $I_C(t) = I_o e^{-t/\tau}$ with $I_o = V_o/R$. (1) Use *Logger Pro* to fit the voltage curves during the charging process and extract from these fits values for the capacitance. Show all your work and reasoning. (2) Integrate the expression for $I_C(t)$ over the interval $t = 0$ to $t = \infty$ and formally prove that it reduces to CV_o. Show all steps.

Before you leave: Make sure your lab table is clean and ready for the next group. Leave equipment neatly organized for the next lab. Please don't leave random scraps of paper, calculators, water-bottles, etc., behind. Use recycling bins down the hall, not the trash, for recyclable materials.

19—Measurements with an oscilloscope

Object

Learn to make basic measurements of voltage and period using an oscilloscope.

Background

This lab is an introduction to the use of the oscilloscope to measure time-dependent electrical signals. The oscilloscope (known to long-time users as simply "the scope") is a widely used laboratory instrument for observing time-dependent electrical signals. It is, in essence, electronic "graph paper." While some may regard it as the domain of the electrical engineer or mad physicist, it finds wide use in a variety of engineering and laboratory settings wherever events occur too rapidly for the human hand to record or the eye to follow without some assistance.

The first encounter with an oscilloscope is often an intimidating event. Oscilloscopes first seem to have an infinite number of controls. In most circumstances the controls you need to use are few in number. Most people who use oscilloscopes regularly become familiar with the details of the equipment only with extensive use and practice. Be patient and carefully watch the effects of each control as you (often desperately) adjust the various knobs and switches. Remember, even experienced scope users encounter difficulties when faced with a new model. Just finding the power switch can be a challenge.

Modern digital oscilloscopes measure the input signals at regular time intervals (the sampling rate) and pass those values on to the display electronics, so what one sees on the display is a snapshot of the signal. Important controls are grouped by function. The two most important groups are the horizontal controls and the vertical controls. The horizontal controls, or timebase settings, specify what total time interval the snapshot of the signal represents, or how much time one horizontal division on the screen represents. The vertical controls set the voltage scale for measured signals: how many volts one vertical division represents. The most common adjustments are made using the control knobs in the respective sections of the front panel. More options are provided through on-screen menus. A two-channel scope is capable of displaying two different input signals simultaneously.

While the scope has some fancy built-in measurement capabilities, all measurements must be done by examining the displayed waveform and applying the horizontal and vertical scale settings to find amplitudes, periods, etc. For this lab we use just the basic features and build up the ability to interpret the waveforms in terms of the horizontal and vertical scales factors, not more advanced tools in the scope.

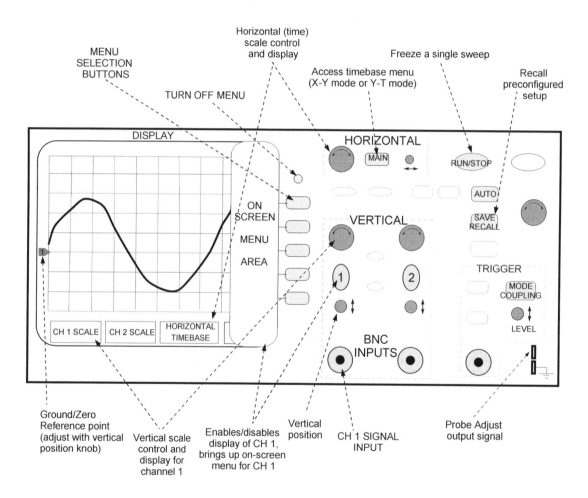

Figure 1: Main features on front panel of a digital oscilloscope

Important settings of the oscilloscope are presented at the bottom of the display area. In particular there will be the vertical scales or sensitivities (volts per division of the grid) for each channel and the horizontal or time base scale (in seconds per division). Other settings are changed by pressing one of the keys and working through on-screen menus.

Display of signals applied to channel 1 or channel 2 can be enabled and disabled by pressing the corresponding numbered keys in the vertical controls section. The vertical sensitivity or scale factor (volts/division) is adjusted by the knob above the numbered key. The horizontal scale or timebase setting (seconds/division) is adjusted using the knob on the left of horizontal controls section.

The TRIGGER section is the least understood aspect for beginners; the default settings are generally appropriate; usually the only adjustment that needs to be made is to the trigger level via the small knob in the trigger controls.

The RUN/STOP button provides a way to freeze the display with the most recently collected data. This is useful if the signal is noisy or the display seems "jumpy" and hard to see a good reproducible result.

On-screen menus can be closed and the full signal display restored using the small round button above the menu selection keys at the immediate right of the display.

AUTO Pressing this key causes the scope to adjust its horizontal and vertical settings to "optimal" values for the input signals. Sometimes this works nicely; often it leads to settings that are completely inappropriate for the signals you are studying. If you use this button, you must always look at the vertical and horizontal settings at the bottom of the display area to see what the scope has selected, and then ask yourself if they are reasonable for what you are doing. If you use the AUTO key and don't see anything that makes sense, try returning to a known set of reasonable choices by recalling a pre-set configuration:

Press SAVE/RECALL

Select STORAGE SETUPS on the on-screen menu

Make sure that Setup no. 1 is selected, Press the LOAD menu key, followed by menu off button.

This restores settings appropriate to the start of the experiment. Make adjustments manually from this point.

Procedure

Basic controls. The first step in becoming familiar with the oscilloscope is to use it to observe a known signal and observe the effects of various controls. The oscilloscope provides an internally generated signal. Sometimes its labeled as "Probe Adjust," or "Cal" but often may simply have some obscure icon or legend indicating the amplitude of the square-wave signal provided. This is normally used to calibrate special adapters used with the scope. For our purposes it provides a convenient signal with known frequency and amplitude.

Frequency is defined to be the number of cycles completed in 1 second and is routinely given the unit of hertz (Hz). A frequency of 60 Hz means 60 complete cycles are executed in one second. (The archaic but more descriptive designation for this measure is cps—for cycles per second.) With oscilloscopes one measures the period of a signal: the time required to complete one full cycle, and then uses the relation between frequency f and period T to find the frequency of the signal: $f = 1/T$. The signal provided on the front of the scope is labeled as being (approximately) 3 V amplitude and has a frequency of about 1000 Hz. This provides a known signal to get familiar with interpreting the horizontal and vertical scale settings of the scope.

Use a coaxial cable with a BNC connector on one end to connect to the scope's channel 1 input and an alligator clip to connect to the upper PROBE ADJUST contact. The convention with BNC cables is that the red lead is the inner conductor of the coaxial cable and BNC connector, while the black lead is the outer conductor, which is to be connected to the electrical ground. The outer conductor of the BNC input of the scope is tied to the scope's ground internally. The alligator-clip end of the ground wire (black) can be left floating (unconnected) for this part. Hook up the signal from the PROBE ADJ to channel 1.

Observe the pattern displayed on the scope. Try adjusting the vertical sensitivity and note the results. Try adjusting the horizontal (time-base) control. For time base settings of 0.1 ms/div (=100μs/div) and 0.5 ms/div (=500μs/div) and vertical sensitivities of 1.0V/div and 2.0 V/div (4 combinations) make careful sketches of the waveform displayed. Adjust the trigger level control to achieve a stable display if necessary. Verify the frequency by determining the period of the waveform from the display and calculating $f = 1/T$. Show explicitly how you calculate frequency and amplitude from the displayed waveform for the two cases (0.1 ms/div, 1V/div) and (0.5ms/div, 2V/div).

Determination of an "unknown" frequency. Sound is a periodic variation in the pressure and density of air (or other gases or liquids). A tuning fork produces a fairly pure sinusoidal pressure variation (if not struck overly hard) at a well-defined frequency. A microphone or speaker can be used to convert the pressure variations of the sound wave into a varying voltage that can be measured with an oscilloscope. Devices such as the speaker or microphone that convert mechanical variations into electrical variations are called *transducers*.

Use the scope and the microphone and amplifier box to observe the sound wave and determine the frequency of the tuning forks. The RUN/STOP button is a handy way to capture the waveform for closer inspection. To achieve the greatest precision in finding the frequency $f = 1/T$, does it make any difference to set the scope to display approximately one complete cycle across the screen and measure the period directly for one cycle, or set the scope to display several complete cycles and measure the time for N cycles, dividing the measured time by N to find the period of one cycle? In addition, is it better to choose as a reference point for your measurements the top (or bottom) of the waveform or the point where the waveform crosses zero? Justify your choice of method.

Compare the frequencies you have measured with those stamped on the tuning forks. Make a sketch to scale of the observed waveform for one tuning fork.

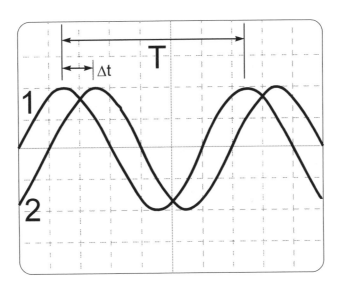

Figure 2: Two signals differing in phase.

Phase measurements: Two signals may have the same frequency, but reach their maxima (or minima) at different times. Such signals are said to differ in phase. Two sinusoidal signals of the same frequency may be written, in general form as $A_1 \sin(\omega t + \phi_1)$ and $A_2 \sin(\omega t + \phi_2)$. The A_1 and A_2 are the amplitudes of the signals, and $\omega = 2\pi f$ is the angular frequency. ϕ_1 and ϕ_2 are the phases. The phase determines the value of each signal at $t = 0$. It is an angle and can be commonly expressed in radians or in degrees. Since the oscilloscope allows us to choose an arbitrary instant as $t = 0$, it's possible to set $\phi_1 = 0$. (Choosing the phase is equivalent to shifting the graph of the function along the time axis.) Then it's possible to re-write the two signals as $A_1 \sin(\omega t)$ and $A_2 \sin(\omega t + \phi)$. Where ϕ is called the phase difference between the two signals. If $\phi > 0$ signal 2 will appear to reach a maximum before signal 1; signal 2 is said to "lead" signal 1. If $\phi < 0$, signal 2 will appear to reach a maximum after signal 1; signal 2 then "lags" signal 1 in time. Note that it is meaningful to refer to phases only over a limited range. Usually the range of the phase shift is restricted either to $-\pi < \phi < \pi$ or $0 < \phi < 2\pi$ radians or equivalently $-180 < \phi < 180$ or $0 < \phi < 360$ degrees. The range used is a matter of choice. (To understand why the range is limited, consider what happens to the appearance of the two signals if $\phi = 2\pi$ radians exactly.) Most oscilloscopes are capable of displaying two signals simultaneously. This makes measurements of phase differences relatively painless. Figure 3 shows a typical display of two signals out of phase. The period of the waveforms is T and it's apparent that signal 2 lags signal 1 by an amount of time Δt. The ratio $\Delta t/T$ is then the fraction of one complete cycle by which signal 2 lags signal 1. The magnitude of the phase shift is then either $\frac{\Delta t}{T}2\pi$ radians or $\frac{\Delta t}{T}360$ degrees.

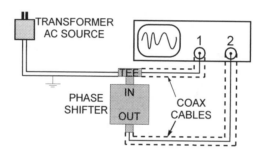

Figure 3: Circuit for phase shift measurements.

You have an AC voltage source in the form of a small transformer that plugs into a standard wall socket and a phase shifter "black box"[1] that will phase shift the signal. Note: AC (i.e., alternating current) is widely used as a generic term for time-dependent signals, even if one is measuring voltages, not currents. Use the provided connectors (BNC tees, coaxial cables, etc.) to connect the voltage source to both the small metal box's input BNC connector and channel 1 of the scope. Use channel 2 of the scope to observe the output of the box. See Fig. 3.

(a) Measure and report the amplitudes and frequency(ies) of the two signals.

(b) Make a careful sketch of the display from which you make your measurements.

(c) Measure the phase of channel 2 relative to channel 1. Clearly display your calculations. Does signal 2 lead or lag signal 1?

---

[1]A "black box" is a box that contains unknown components. In fact the box you will work with is probably dull metallic in appearance, not actually black.

X-Y mode and Lissajous figures. No introduction to the oscilloscope would be complete without some exotic (and often confusing) displays in the form of Lissajous figures. For this part, the horizontal axis of the scope is controlled by the voltage applied to channel 1 (X) and the vertical axis is controlled by the input to channel 2 (Y).

Getting into X-Y mode (CH1 determines X and CH2 determines Y) requires accessing an on-screen menu. Press the **MAIN/DELAYED** key in the Horizontal section of the controls to bring up the on-screen menu for time-base settings. Then press the **Time Base** menu key until it displays **Time Base X-Y**. Follow the same procedure to restore normal operation via **Time Base Y-T**. NOTE: You will need to adjust the timebase setting (seconds/div) whenever coming back to **Y-T** mode from **X-Y**.

To achieve a nice Lissajous figure (as seen in the mad scientist's lab in old sci-fi movies) once X-Y mode is established, use the horizontal control knob to adjust the scope's sampling rate to ≤ 200 kSa/s (200,000 Samples per second or a sampling frequency of 200 kHz). This will be displayed in the lower right of the screen replacing the normal timebase setting display.

To begin, use the AC voltage source from the previous part and a BNC "tee" to connect the same signal to both channels 1 and 2. (Disconnect the tee from the phase shifter box and move it to channel 1 (X) and connect a coax cable from the tee to channel 2 (Y). The transformer should remain connected to the tee at the channel 1 input.) Make sure that both vertical sensitivity controls are set to the same value. The scope should reveal a straight line, with a slope of 1, since the scope is doing nothing more than plotting the function $Y = X$ repeatedly. Try inverting channel 2. (This takes the signal at CH2 and multiplies it by -1 inside the scope; press the CH2 button and from the menu turn "Invert" on.) Verify that the slope of the line becomes -1 ($Y = -X$). Try adjusting the vertical sensitivity controls to verify you understand what the scope is doing. This display is not very interesting because the two input signals, X and Y, are at the same frequency and precisely in phase. Turn the CH2 "Invert" off.

Next, leaving channel 1 (X) as it is (the AC signal straight in), replace the Y input by feeding the AC signal into the phase shifter "black box" and take the output of this box and connect it to Y. (Use the same circuit as Fig. 3, except the scope is in X-Y mode.) You may need to crank up the sensitivity on Y to get a proper display because the box attenuates the input signal. You should now see a stable elliptical display as a consequence of the phase and amplitude changes caused by the black box. Again try inverting channel 2 to observe the results. Sketch your observations.

Hook-up the output of the function generator to channel 2. The function generator provides a signal of variable frequency and amplitude. Switch the scope out of X-Y mode and adjust the amplitude and frequency controls of the generator to produce a signal of about 60 Hz, with approximately the same amplitude as the AC signal attached to channel 1. Once this is done, hop back to X-Y mode. Slight adjustments of the frequency of the function generator should produce a pattern that seems to "rotate" slowly. Adjust the function generator to make the pattern change as slowly as possible and sketch the pattern. (RUN/STOP is helpful in freezing the display, again.) Then try frequencies of 30 and 120 Hz on the function generator. Again adjust the frequency control to achieve a (nearly) stationary display. *Sketch for each case.* Slowly turn up the frequency and observe at what frequencies the apparent "rotation" of the display almost stops. Explain why these certain frequencies produce a stationary pattern.

Before you leave: Restore your scope to a reasonable set of default settings for the next lab:
Press SAVE/RECALL
Select STORAGE SETUPS on the on-screen menu
Make sure that Setup no. 1 is selected, Press the LOAD menu key, followed by menu off button.
This restores settings appropriate to the start of the experiment.

Make sure your lab table is clean and ready for the next group. Leave equipment neatly organized for the next lab. Please don't leave random scraps of paper, calculators, water-bottles, etc., behind. Use recycling bins down the hall, not the trash, for recyclable materials.

20—Current balance

Object

To measure the magnetic force between parallel electric currents and its dependence on the currents and distance.

Background

A charged particle moving through a magnetic field is subject to a force given by $\vec{F} = q\vec{v} \times \vec{B}$. In this lab, the magnetic field is produced by an electric current I_1 flowing through a straight length of a fixed conductor. The magnetic field produced by this current acts not on a single charge, but on a second current, I_2, flowing through a parallel conductor that is free to move in response to the magnetic force. For parallel conductors with the currents flowing in the opposite directions the force is repulsive. Its magnitude is

$$F_m = \frac{\mu_o}{2\pi} \frac{I_1 I_2 L}{r}, \tag{1}$$

where r is the distance between the conductors, L is their common length, and μ_o, the magnetic permeability of free space, $= 4\pi \times 10^{-7} N/A^2$. The situation considered here will be one in which the same current is routed through both conductors in series, so that $I_1 = I_2 = I$ and the force is then

$$F_m = \frac{\mu_o}{2\pi} \frac{L}{r} I^2. \tag{2}$$

The goal is to verify that the force is proportional to I^2 and inversely proportional to r through graphical analysis of various sets of data.

The apparatus consists of a fixed conductor (with diameter, $D = 0.318$ cm) and a second parallel conductor mounted on a knife-edge (K) so that it may pivot. The equilibrium separation, r, of the two conductors is determined by the position of an adjustable counterweight (CW). See Fig. 1. This is a delicate apparatus and should be treated accordingly. The two conductors must remain carefully aligned and as parallel as possible during

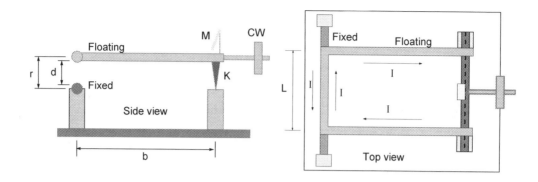

Figure 1: The current balance measures the force between parallel currents.

the measurements. Handle them gently and as infrequently as possible.

The floating conductor can be displaced by adding a small mass, m, to a small pan attached to it. The weight of the added mass provides a torque (mgb) about the pivot point, causing a new equilibrium position for the floating conductor, perhaps coming to rest against the fixed conductor. By applying current to the system so that there is a repulsive magnetic force $\vec{F}_m$ between the conductors, another torque is produced that can be adjusted to restore the floating bar to its original equilibrium position. The magnitude of $\vec{F}_m$ must equal the weight of the added mass for the torques to cancel. A series of masses can be added and the resulting current required to restore the original equilibrium recorded. The separation r between the centers of the conductors is not easily measured directly because it is usually just a few mm and the system is easily disturbed. Instead, the distance between the outside of the conductors (d) and their diameter ($D = 0.318\,\text{cm}$) are combined to find $r = d + D$.

The measurement of d is made with a laser and a long lever arm to amplify the value of d. By adding sufficient mass to bring the conductors together, the mirror, M, tilts and the laser beam is displaced on the screen by an easily measured amount as shown in Fig. 2.

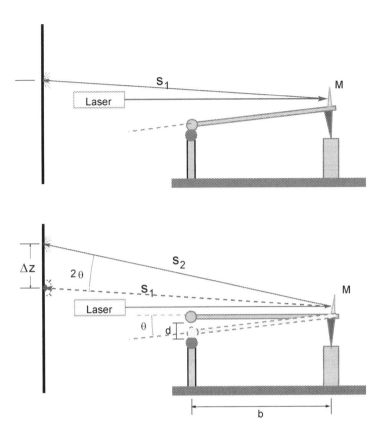

Figure 2: A laser-based optical lever amplifies the displacement of the upper conductor, displacing the laser light spot Δz when the conductor moves a smaller distance d. Notice that the rotation of the upper conductor by θ causes a 2θ change in direction of the reflected laser beam.

Measuring the appropriate lengths and a bit of geometry will provide a measure of d. This is illustrated in Fig. 2. The distances s_1, s_2, and Δz and angle 2θ can be related by the law of cosines,

$$(\Delta z)^2 = s_1^2 + s_2^2 - 2s_1 s_2 \cos(2\theta),$$

so that

$$\theta = \frac{1}{2} \cos^{-1} \left(\frac{s_1^2 + s_2^2 - (\Delta z)^2}{2s_1 s_2} \right).$$

In practice, the angles may be smaller than shown, so that $s_1 \approx s_2$ and small angle approximations, such as $\tan 2\theta \approx 2\theta$, so that $2\theta \approx \frac{\Delta z}{s_1}$ may be valid to obtain a simple result for d in terms of s and the displacement Δz of the laser spot on the screen. Examine your situation carefully to determine if these simplifying approximations can be used.

Procedure

The current through the system is provided
by a power supply. Include a 0.4 Ω resistor
(R_s) in series with the apparatus. The current
will be found by measuring the voltage drop
across this resistor and applying Ohm's law
$I = V/R_s$. Wire up the system so currents will
flow in opposite directions through the parallel
segments of the conductors and through R_s.

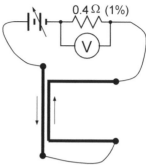

Figure.3:

Adjust the counterweight (CW) to produce a distance between the conductors, d, of 1–2 mm. Make certain the conductors are aligned when viewed
from above and from the side. Mark on the screen the equilibrium location
of the laser beam spot without any added masses. Add a 500 mg mass to the
floating conductor (gently, using tweezers) to bring the floating conductor to
rest against the fixed conductor. Measure the distance the laser beam has
been displaced from the original equilibrium spot. Also note the distance(s)
from the mirror to the screen. This allows you to find 2θ. With an additional measurement of b on the apparatus, you have enough information to
determine d and then r, the center-to-center distance between the parallel
conductors. (For small θ, $d \approx b\theta$, provided you measure angles in radians!)
Carefully remove the 500 mg mass and make sure the original equilibrium
position is restored. You may want to examine how close the conductors are
to being parallel by looking for gaps when they are in contact.

AT NO TIME DURING THE MEASUREMENTS SHOULD THE CURRENT
EXCEED 15 A! DO NOT ALLOW CURRENTS GREATER THAN 10 A TO
RUN CONTINUOUSLY FOR MORE THAN 60 SECONDS. EXCESSIVE
HEATING CAN CAUSE THE CONDUCTORS TO EXPAND AND DISTORT,
SPOILING THE DATA AND POSSIBLY DAMAGING THE APPARATUS.
ALWAYS REDUCE THE CURRENT IMMEDIATELY AFTER A
MEASUREMENT IS MADE.

For $R_s = 0.4\,\Omega$, the voltage drop across R_s must *never* exceed 15 A $\times$
$0.4\,\Omega = 6.0\,\text{V}$ and should not be allowed to stay at $\geq 4\,\text{V}$ for more than 60
seconds.

THIS RESISTOR DISSIPATES SIGNIFICANT POWER AND CAN GET *hot*!
DO NOT TOUCH IT!

Next add a series of four or five masses—no more than 200 mg and this much only if you manage to adjust the counterweight to produce a very small equilibrium separation. For each mass increase the current until the laser beam is restored to its unloaded equilibrium position. Record the current needed to bring the laser beam back to the unloaded equilibrium spot. This state corresponds to the magnetic force balancing the added weight: $F_m = mg$.

Change the position of the counterweight. Determine the new equilibrium laser spot position, the new d and the new equilibrium separation r between the centers of the conductors. Repeat the measurements using a series of 4-5 masses with this new r.

Collect data sets for a total of four different positions of the counterweight, each with a different r.

Re-wire the circuit so that the current flows in the same direction through the two conductors. Qualitatively verify that the magnetic force between the currents is now attractive. Describe what you observe that demonstrates this.

Analysis. According to Eq. 2, a plot of F_m as a function of I^2 should be a straight line for constant r. For each set of data collected corresponding to a particular equilibrium r fixed by a particular counterweight position, plot the forces ($= mg$) as a function of I^2. The resulting plots should be four straight lines if Eq. 2 is valid. Evaluate the slope of the best fit line for each set, which should, according to Eq. 2, be $\mu_o L/2\pi r$. Note in drawing the lines that according to Eq. 2, the lines are expected to pass through the origin since the model predicts $F = 0$ when $I = 0$. Superimpose all four data sets (and four best-fit lines) on one graph, making each data set easily distinguishable by using a different symbol to plot each set.

Does your graph support the claim of Eq. 2 that the magnetic force $F_m \propto I^2$?

Distance dependence in Eq. 2. According to Eq. 2, the slopes of the four lines plotted above are expected to be

$$\text{slope} = \frac{\mu_o L}{2\pi} \frac{1}{r}.$$

This can be tested. Make a table of $1/r$ and the slopes just found from the fitted lines for the four sets of data. Then use these data to make another plot of the slopes a function of $1/r$.

Do these data fall along a straight line? If so, draw the best fit straight line through these data and compare the slope of *this* line to the predicted value $\frac{\mu_o L}{2\pi}$ (Be sure to measure L, the common length of the parallel conductors. See Fig. 1 again.) **Pay close attention to units throughout your calculations.**

Are your overall results in agreement with the model, Eq. 2?

Before you leave: Make sure your lab table is clean and ready for the next group. Leave equipment neatly organized for the next lab. Please don't leave random scraps of paper, calculators, water-bottles, etc., behind. Use recycling bins down the hall, not the trash, for recyclable materials.

21—Earth's magnetic field

Object

To measure the horizontal component of Earth's local magnetic field and determine the magnetic moment of a permanent magnet.

Background

Earth's magnetic field has both a horizontal component and a vertical component. This experiment measures the horizontal component only, which is denoted here by $\vec{B}_e$. The direction of B_e is toward magnetic north. Both the direction and value of the local $\vec{B}_e$ will be affected by iron and other magnetic materials in the building frame and furniture. Measurement of $\vec{B}_e$ will be done by the deflection or tangent magnetometer method, using Helmholtz coils to create a known magnetic field $\vec{B}_h$. $\vec{B}_h$ is arranged to be horizontal and perpendicular $\vec{B}_e$.

Helmholtz coils are a set of identical coils of wire carrying a current. To calculate the strength of the field, B_h, produced by the Helmholtz coils, refer first to the formula for the magnetic field at a point on the axis of one circular turn of wire with radius R at distance x from the center of the coil, $B_x(x) = \frac{\mu_o I R^2}{2(R^2+x^2)^{3/2}}$. I is the current flowing through the wire.

Helmholtz recognized that a pair of identical coils separated by a distance equal to their radius produces a field midway between the coils that is especially uniform. It is straightforward to show

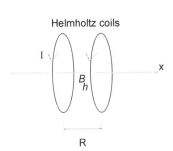

Figure 1:

that the field along the axis midway between the centers of both coils is

$$B_h = \frac{8\mu_o N}{\sqrt{125}R}I = C_h I \,. \tag{1}$$

R is the average radius of all turns of wire on each coil; N is the number of turns in one coil, which is same for both coils. $C_h = \frac{8\mu_o N}{\sqrt{125}R}$ is a convenient

155

proportionality factor between the current and the B field you will calculate
for your coils.

Let $\vec{B}_r$ be the resultant of the two horizontal magnetic fields from Earth
and the Helmholtz coils: $\vec{B}_e$ and $\vec{B}_h$. Let α be the angle between this resultant
field, $\vec{B}_r$, and the horizontal Earth's field $\vec{B}_e$ as in Fig. 2. Then

$$\tan \alpha = \frac{B_h}{B_e}.$$

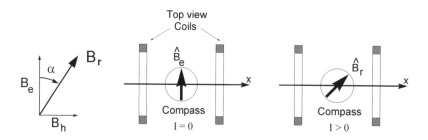

Figure 2: A compass shows the direction of the net horizontal magnetic field.

When the current is off in the Helmholtz coils, $\vec{B}_h$ is zero so that $\alpha = 0$ and
a magnetic compass would indicate the direction of $\vec{B}_e$. When the current
is on, a magnetic compass would indicate the direction of $\vec{B}_r$, so that by
knowing α and the value of $\vec{B}_h$ one can calculate $\vec{B}_e$.

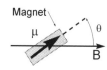

Figure 3: A restoring
torque acts on a mag-
net not aligned with
the B field.

A permanent magnet can be characterized by a
magnetic (dipole) moment, $\vec{\mu}$. A magnetic moment
in a magnetic field experiences a torque when not
aligned with the local magnetic field: $\vec{\tau} = \vec{\mu} \times \vec{B}$.

For the situation shown in the figure, $\tau = -\mu B \sin \theta$. The minus sign indicates the torque tends
to restore $\vec{\mu}$ to the direction of $\vec{B}$, and θ is measured
from the direction of $\vec{B}$. Since torque can be related
back to angular momentum $\vec{L}$ using the rotational
version of Newton's 2nd law for the magnet:

$$\vec{\tau} = \frac{d\vec{L}}{dt}$$

$$\tau = I_m \frac{d^2\theta}{dt^2}$$

where I_m is the rotational inertia or moment of inertia of the magnet. Then the deflection of the magnet, θ, is described by

$$
\begin{aligned}
I_m \frac{d^2\theta}{dt^2} &= -\mu B \sin\theta \\
&\approx -\mu B \theta
\end{aligned}
$$

since for small angles $\sin\theta \approx \theta$. This gives $\frac{d^2\theta}{dt^2} = -\frac{\mu B}{I_m}\theta$. The result is simple harmonic motion as the magnet oscillates with $\theta(t) = \theta_o \cos(\omega t)$ about the equilibrium direction with an angular frequency, $\omega = (\frac{\mu B}{I_m})^{1/2}$. This is mathematically analogous to a mass on a spring where $\frac{d^2x}{dt^2} = -\frac{k}{m}x$ and $\omega = \sqrt{\frac{k}{m}}$. The magnet oscillates more rapidly in a stronger B field. The period, T, or time for one complete cycle of oscillation gets shorter in a stronger B field and is

$$
T = \frac{2\pi}{\omega} = 2\pi\sqrt{\frac{I_m}{\mu B}}.
$$

This can be cast into the form

$$
\frac{1}{T^2} = \frac{\mu}{4\pi^2 I_m} B, \tag{2}
$$

which predicts a graph of T^{-2} as a function of magnetic field will be a straight line. A knowledge of B, T, and I_m allows the magnitude of the magnetic moment $|\vec{\mu}|$ to be determined.

Procedure

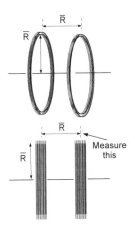

Figure 4: Finding the average radius of the coil's turns.

Preliminaries. Record all the information required to calculate the magnetic field that will be produced by your Helmholtz coil pair. The number of turns in one of the coils should be marked on the apparatus.

You will need the average radius R of the coils. Let's assume that the coils are properly designed, so the average radius should be equal to the distance along the axis between the centers of the coil windings. Measure this distance and use this value to calculate the constant C_h. You might find it convenient to express the value for C_h for your Helmholtz coils in units of $\mu\text{T/mA}$ (microtesla per milliamp) since the B fields here are typically 10–20 μT and the currents needed in the coils are usually measured in mA.

Measuring the local B_e. Connect the circuit as shown in Fig. 5. Lay out the circuit so that the Helmholtz coils are not close to power supply or current meter. Use long leads, preferably twisted together, between the Helmholtz coils and the switch. The double-pole-double-throw (DPDT) switch is wired so the current in Helmholtz coils can be reversed easily (without reversing the direction of the current through the ammeter). Have your lab instructor check your circuit before applying power. Place the magnetic compass on

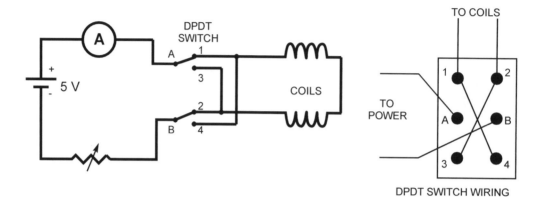

Figure 5:

the post in the center of the Helmholtz coils. Carefully align the coils so that the axis of the coils is perpendicular to the direction of compass needle. This will ensure that B_h is perpendicular to B_e.

With a large resistance (~ 5 kΩ) set on the decade resistance box, turn on the current by closing the switch. Slowly reduce the resistance of the decade box, increasing current, until the compass has deflected to an angle, α, of 45°. Record the current and angle. Reverse the current direction with the switch and read the deflection angle in the other direction. Use the average of these two values as α_{av} in your calculations.

Repeat this process at three other values of current that will give deflections of about 25°, 35° and 55°. In each case, record the current and the angle of deflection for both directions of current flow. Record data in a neatly organized table, with one line for each current value used.

In the calculations of magnetic field, you should find and list in your table, for each value of the current used, the average angle α_{av}, the value of $B_h = C_h I$, and the value of $B_e = B_h / \tan(\alpha_{av})$. Ideally, the four values for B_e listed in the last column should all be the same if the experiment were free of errors. Find the average of the four values of B_e. Take this average as your final result for the determination of B_e. Report your result in units of μT (microtesla). It is expected that B_e should be $\sim 10 - 20$ μT. The exact value is very sensitive to the presence of iron or steel in the vicinity, such as in the lab table. The values can differ noticeably from table to table and even with position on any one table.

Measuring the magnetic moment of a magnet. Without moving the center of the apparatus, rotate the Helmholtz coils 90° so that $\vec{B}_h$ is parallel to $\vec{B}_e$. Place the suspended permanent magnet between the coils. The magnet will oscillate with a period determined by the net magnetic field, $\vec{B}_{net} = \vec{B}_e + \vec{B}_h$.

Determine the period of oscillation, T, for the magnet using at least four different magnetic fields, ranging from $|\vec{B}_{net}| \approx 5\,\mu\text{T}$ to $|\vec{B}_{net}| \approx 40\,\mu\text{T}$. You can start the oscillations by momentarily flipping the current direction with the switch. Measure the time required for 10 or 20 oscillations in order to accurately measure T. Pay attention to whether or not the $\vec{B}_h$ you use is in the same or opposite direction to $\vec{B}_e$ when you determine the net magnetic field $\vec{B}_{net}$.

Make measurements of the magnet's mass and dimensions to allow an estimation of its moment of inertia. The moment of inertia of the magnet is that of a solid cylinder rotating about an axis passing through its center and perpendicular to its own length: $I_m = \frac{1}{4}MR_m^2 + \frac{1}{12}ML_m^2$, where M is its mass, R_m is the magnet's radius (0.318 cm) and L_m is the magnet's length.

Make a graph of $1/T^2$ as a function of B. According to Eq. 2, this plot should be a straight line that passes through the origin. Using the slope of this line and your other measurements of the magnet, calculate the value of $|\vec{\mu}|$ for the magnet from the slope of the graph. Pay close attention to the units. Do not confuse the magnitude of the magnetic moment, $\mu = |\vec{\mu}|$ with the magnet's mass M.

Before you leave: Make sure your lab table is clean and ready for the next group. Leave equipment neatly organized for the next lab. Please don't leave random scraps of paper, calculators, water-bottles, etc., behind. Use recycling bins down the hall, not the trash, for recyclable materials.

Questions

(1) Earth's magnetic field has both horizontal and vertical components. A dip needle is a compass that has been carefully balanced so that you can rotate it out of the usual horizontal plane and use it to sense the vertical component of B. Use the dip needle set up in lab to observe the angle the Earth's magnetic field makes with the horizontal plane. This angle is typically around 70° locally—Earth's magnetic field is mostly directed down into the ground at our location. The ratio of vertical and horizontal components can again be expressed as the tangent of an angle. Use your measured hor-

izontal B_e and the dip angle to calculate the vertical component of B and then magnitude of the total local B.

(2) A single loop of wire with area A carrying a current I has a magnetic moment $\mu = IA$. If you wanted a single loop of wire with an area equal to the cross-sectional area of the magnet used here to have the same magnetic moment as your permanent magnet, how large a current would you need flowing through the loop?

(3) Adapt the expression for the magnetic field due to one turn of the coil given earlier, $B(x) = \frac{\mu_o I R^2}{2(R^2 + x^2)^{3/2}}$, to the situation of two coils, one each at $x = \pm R/2$, and derive the Helmholtz coil result given in Eq. 1. (Hint: putting a coil at $x = \pm\frac{R}{2}$ means replacing x by $x \pm \frac{R}{2}$, add the fields from both coils and evaluate at $x = 0$.)

22—Magnetic induction – Faraday's law

Object

To become familiar with magnetic induction, Faraday's law, and Lenz's law through a series of investigations using a permanent magnet and wire coils.

Background

Magnetic induction refers to several phenomena, including those where a changing magnetic field induces an electric field and thereby a current or a voltage in a nearby conductor. For example, waving a permanent magnet near a loop of wire will induce an electric current in the loop. This induced current itself creates a magnetic field. The induced magnetic field tends to keep the total magnetic flux through the loop constant—the induced current flows so as to oppose the change in magnetic field that causes it. This is the essence of Lenz's law.

If the loop is opened and a voltmeter attached, the induced voltage or electromotive force (EMF) that will cause the current to flow can be measured. Faraday's law relates the rate of change of magnetic flux to the induced voltage drop around the loop:

$$V_{ind} = \oint \vec{E} \cdot \vec{dl} = -\frac{d\Phi}{dt}. \qquad (1)$$

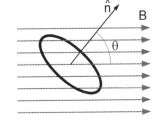

Figure 1: Loop in uniform field.

The magnetic flux is determined by integrating the component of the magnetic field normal to the loop over the area of the loop $\Phi = \int \vec{B} \cdot \hat{n}\, dA$.

For spatially uniform (but still time-dependent) magnetic fields, the magnetic flux through one loop of wire is given simply by $\Phi = \vec{B} \cdot \hat{n} A = BA\cos\theta$ where $\vec{B}$ is the magnetic field strength in tesla (T) and A is the area of the loop. θ is the angle between the $\vec{B}$ field and the vector $\hat{n}$, normal or perpendicular to the plane of the loop. If no magnetic field lines cut through the area of the loop ($\theta = 90°$), the magnetic flux is zero. When the single loop is replaced by a coil consisting of N_d turns, the flux through the coil is N_d times larger. If the magnetic field oscillates at a frequency f, with angular frequency $\omega = 2\pi f$, so that $\vec{B}(t) = \vec{B}_o \cos(\omega t)$, the magnetic flux will have the same time dependence and the induced voltage will then be

$$V_{ind}(t) = -N_d A \cos\theta \frac{\partial B}{\partial t} = \omega N_d A B_o \cos\theta \sin(\omega t). \tag{2}$$

This induced voltage oscillates at the same frequency as the magnetic field, but one quarter cycle out of phase, $V_{ind}(t) = V_o \sin(\omega t)$ with the amplitude V_o given by

$$V_o = \omega N_d A B_o \cos\theta. \tag{3}$$

This result can be used to find the amplitude of an oscillatory magnetic field by solving the previous equation for B_o, and measuring the amplitude of the oscillating induced voltage in the detector coil:

$$B_o = \frac{V_o}{\omega N_d A \cos\theta}. \tag{4}$$

In this lab you will first carry out a set of qualitative observations of the current induced in a small detector coil when a permanent magnet is moved. Then you will quantitatively measure the induced voltage and verify various aspects of Faraday's law by comparing the measured induced voltage to that predicted in Eq. 3 from Faraday's law.

Procedure

Qualitative observations. A permanent bar magnet with poles labeled N and S is provided, along with a galvanometer and two detector or pick-up coils on long handles. The galvanometer is a sensitive current-measuring device (an ammeter), which has three ranges of increasing sensitivity and shows the direction of the current by the direction of deflection of its needle. A positive current (i.e., entering the meter at the + terminal and flowing out the − terminal) causes a positive (rightward) deflection. Currents flowing in the opposite sense will cause an opposite deflection. To use the galvanometer one of the three range buttons must be pressed (and held) to complete the circuit. Always begin measuring with the least sensitive range. If little or no deflection is observed in your measurements, try successively more sensitive ranges.

Begin with the detector coil consisting of 400 turns of copper wire mounted on the end of a plastic handle and two sockets for banana plug cables. Connect the detector coil and galvanometer in a simple circuit as shown. Pay careful attention to the legend on the detector coil showing the direction of winding of the wire in the coil.

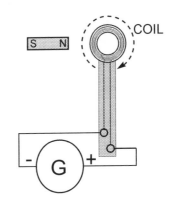

Figure 2: Detector coil and galvanometer.

Wave the permanent magnet near the detector coil. (Be sure to begin with the least sensitive galvanometer scale.) Try moving the magnet from side to side and also toward and away from the detector coil. Try moving the magnet at different speeds. What kinds of motion induce the largest current in the coil? What happens if the magnet is held fixed and the detector coil is moved instead?

Before proceeding any further, discuss among your lab team whether or not your observations are qualitatively consistent with the claim that the induced current is proportional to the rate of change of magnetic flux.

Lenz's law. Carry out a more systematic study of the *direction* of the induced current and of the consequent *induced* magnetic field. Begin by making some predictions as outlined below. Record your predictions with supporting diagrams in your notebook *before* you make any measurements.

- As the N pole of the magnet is brought closer to the detector coil (oriented so that it could pass through the central hole of the coil), do you expect the magnetic flux due to the magnet cutting through the detector coil to increase or decrease? Consider these questions in developing your prediction: Do magnetic field lines point away from or toward the N pole of a permanent magnet? How does the magnetic field strength vary with distance?

- According to Lenz's law, the induced current that flows produces a magnetic field that opposes the change in flux from the magnet and tries to keep the total flux (i.e., flux due to magnet + flux due to induced current) constant. In which direction must this induced magnetic field point to do this?

- In which direction must the current flow around the loops of the detector coil to produce this induced magnetic field? (Recall the right-hand rule that relates the direction of the current in a wire to the direction of the magnetic field produced by the current in the wire.)

- Which way should the galvanometer deflect as the N pole approaches your detector coil? (Trace out the flow of current through your circuit carefully to make sure your circuit is wired appropriately.)

Try the experiment and compare the observed galvanometer deflection to your predicted deflection. What should happen as the magnet is pulled away? If your observations disagree with you predictions (1) re-think your predictions, discussing them with your lab instructor if necessary, and (2) verify that you have correctly related the galvanometer deflection to the direction the current flow around the loops of the detector coil, double-checking your wiring carefully.

What will happen if you repeat this process using the S pole of the magnet instead? Verify the expected behavior. If the N pole of the magnet is passed over the detector coil at a fixed height, what should happen to the galvanometer? Try it. Replace the 400-turn detector coil with the 2000-turn coil. What differences are apparent with this detector coil?

Measuring an induced voltage. In this part an electric current through a larger diameter coil (the "field coil") will be used to produce a magnetic field. By driving an alternating current through this coil, the magnetic field will oscillate in strength and direction. A detector coil placed near the field coil will then sense a time-varying magnetic flux, which induces a measurable voltage in the detector coil. The induced voltage in the detector coil will be measured with an oscilloscope and the magnitude of the induced voltage will be compared to the value predicted by Faraday's law though Eq. 4.

Setting up the magnetic field: According to the Biot-Savart law, a single circular loop (radius R) of wire carrying a current I_o produces a magnetic field along its axis of $B(z) = \frac{\mu_o}{4\pi} \frac{2\pi R^2 I_o}{(R^2+z^2)^{3/2}}$ pointed along the axis. If N_f turns of wire are used the field is N_f times larger:

$$B(z) = \frac{\mu_o}{4\pi} \frac{2\pi R^2 N_f I_o}{(R^2 + z^2)^{3/2}}. \qquad (5)$$

At the center of the coil ($z = 0$) the magnetic field strength reduces to a simpler result:

$$B(0) = \frac{\mu_o N_f I_o}{2R}.$$

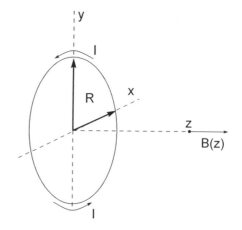

Figure 3: Current carrying loop producing magnetic field.

If the current is time-dependent $I(t) = I_o \cos(\omega t)$, the magnetic field also exhibits exactly the same time dependence. The field coil used here has $N_f = 200$ turns of #22 wire with an average radius of $R = 10.5\,\text{cm}$. A $1.2\,\text{k}\Omega$ resistor, R_s, is in series with the coil. By measuring the voltage drop, V_R, across this resistor, the current can be checked from Ohm's law: $I = V_R/R_s$.

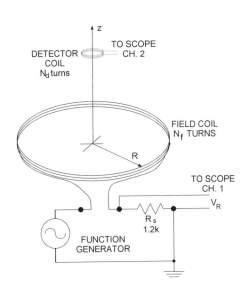

Figure 4: Electrical connections to field coil.

Calculate the current needed to produce a 4.00 μT magnetic field at the center of the field coil. Calculate the corresponding voltage drop V_R needed across the series resistor. Connect the field coil, function generator and oscilloscope in the circuit shown. Set the function generator for a sine wave with a frequency of $f = 500\,\text{Hz}$. Using the oscilloscope to monitor the voltage drop across V_R, adjust the function generator amplitude until V_R oscillates between $\pm$ the calculated value so that the "peak-to-peak" (from minimum to maximum) voltage is twice your calculated value. The magnetic field at the center of the coil will then be oscillating between $\pm 4\,\mu$T: $B(t) = 4.00 \cos(\omega t)\,\mu$T. Since $B \propto I \propto V_R$, the oscilloscope display of $V_R(t)$ is in phase with the magnetic field, $B(t)$. Recall that $\omega = 2\pi f$.

Now when a detector coil positioned at the center of the field coil ($z = 0$ again), it will have a total magnetic flux cutting through it of $\Phi(0, t) = B(0, t) N_d A \cos\theta$, where N_d is the number of turns of wire in the detector coil, A is the average area of the turns of the detector coil and θ is the angle the detector coil's normal vector makes with the z-axis. This assumes that the magnetic field is uniform over the area of the detector coil. If the detector coil is small, this is a reasonable approximation that can be tested experimentally.

Connect the 2000 turn detector coil to the second channel on the oscilloscope and position the detector coil at the center of the field coil with the planes of the coils parallel. This detector coil has an inner diameter of $d_1 = 1.9\,\text{cm}$ and an outer diameter of $d_2 = 3.8\,\text{cm}$. The average area of one turn is $A = \frac{\pi}{3}\frac{(d_1^2 + d_2^2 + d_1 d_2)}{4}$. Observe on the oscilloscope the induced voltage in the detector coil and the magnetic field (via $V_R(t)$) simultaneously. What is the magnetic field when the induced voltage passes through zero? What is the magnetic field when the induced voltage reaches a maximum? What can you say about the phase of the induced voltage relative to the phase of the magnetic field?

Measure the amplitude of the *induced* voltage (V_o = one-half the peak-to-peak value observed on the oscilloscope). Compare this to the value predicted by Eq. 3.

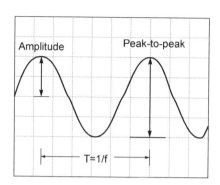

Figure 5: Sinusoidal signal quantities.

Angular dependence. Rotate the detector coil, changing θ, and observe the effect of the orientation of the coil on the induced voltage. Carry out a measurements of the amplitude of the induced voltage as a function of angles of $\theta = 0°$, $30°$, $45°$, $60°$, $90°$. Make a plot of V_o vs. $\cos\theta$. According to Eq. 3 this should be a straight line passing through the origin with a slope (predicted by Eq. 3) of $\omega N_d A B_o$. Are your measurements consistent with this expectation?

Frequency dependence. If the frequency setting of the function generator driving the field coil is doubled to 1.00 kHz, what do you expect to happen to the induced voltage in the detector coil? Carry out the measurement. Also try frequencies of 100, 300, 3000, and 5000 Hz. Record the amplitude of the induced voltage, V_o, for each frequency. **Note:** You may need to adjust the function generator to keep the *amplitude* of $V_R(t)$—and therefore $B(t)$—the same at all frequencies. Make a plot of the amplitude of the induced voltage as a function of frequency for all six frequencies. Is this graph consistent with Eq. 3?

Triangle wave ramp of magnetic field. With the detector coil at the center of the field coil again, change the function generator to produce a triangle wave, so that the magnetic field ramps up and down *linearly* in time. Add a $10\,\text{k}\Omega$ resistor in parallel with the detector coil. (A resistor for this purpose is mounted on a banana plug adapter. This damps out oscillations that may occur in the detector circuit.) Set the function generator frequency to 500 Hz and adjust the amplitude so that $V_R(t)$ is a triangle wave with 8.0 V peak-to-peak. Observe the induced voltage in the detector coil. Make a quantitative sketch of it and explain why it has the particular shape it does.

Variations in the plane of the field coil. Move the detector coil slowly away from the center of the field coil toward the field coil windings. Based on the changes in the induced voltage, how does the strength of the field change? What happens as you move the coil beyond the field coil windings? What happens to the direction of the magnetic field as you move the detector coil past the field coil windings?

Distance dependence along the field coil axis. Use the detector coil to figure out the amplitude of the magnetic field as a function of z along the axis of the field coil with $f = 500\,\mathrm{Hz}$ and the original $4\,\mu\mathrm{T}$ amplitude *sinusoidal* magnetic field at the center. For distances of 5, 10, and 20 cm above the field coil along the z-axis measure the induced voltage and find the magnetic field amplitude $B(z)$ using Eq. 4.

Compare your $B(z)$ values inferred from the induced voltage in the detector coil to those values predicted by Eq. 5. For comparison with expectations, Eq. 5 can be re-written in a more convenient form in terms of the original value at the center of the coil, $B(0) = 4\,\mu\mathrm{T}$:

$$B(z) = \frac{B(0)}{\left[1 + \frac{z^2}{R^2}\right]^{3/2}} \, . \tag{6}$$

Further questions and analysis.

1. There are two possible choices for the direction of $\hat{n}$, the vector normal to the detector coil plane. The magnetic flux depends on the dot product, $\vec{B} \cdot \hat{n}$, so the sign of the flux depends on the direction chosen for $\hat{n}$. Does the choice matter? If so, how do you decide which one to use?

2. Compare the amplitude of the observed induced voltage when a triangle wave was used to drive the field coil to that predicted by Faraday's law. Note that Eq. 4 is valid only for sinusoidal driving currents. To make this comparison, you will first need to back up in the derivation of to consider the question: What is $\frac{dB}{dt}$ for the triangle wave used here to drive the field coil?)

3. The field coil has N_f turns and a finite thickness to it, with turns of radius varying from inner to outer radii, R_i and R_o. Show that when

the varying radii of the turns is accounted for, the magnetic field at the center of the coil is given by

$$B = \frac{\mu_o N_f I}{2(R_o - R_i)} \ln \frac{R_o}{R_i}.$$

Compare the field calculated using this more careful result with Eq. 5 evaluated at $z = 0$ using the average radius of the turns for $R = 10.5\,\text{cm}$. (For the field coils used here the manufacturer claims that $R_i = 9.80\,\text{cm}$ and $R_o = 11.05\,\text{cm}$.)

4. The flux threading the detector coil depends on the average or mean area of the turns. This average area is not merely $\pi \bar{r}^2$, but instead $\pi \overline{r^2}$. Show that the result quoted earlier in the lab for the average area A is the result of calculating $\overline{r^2}$, instead of merely using $\bar{r} = \frac{1}{2}(r_1 + r_2)$. The mean squared value $(\overline{r^2})$ is not the same as the square of the mean $(\bar{r}^2)$!

Before you leave: Make sure your lab table is clean and ready for the next group. Leave equipment neatly organized for the next lab. Please don't leave random scraps of paper, calculators, water-bottles, etc., behind. Use recycling bins down the hall, not the trash, for recyclable materials.

23—Reflection and refraction by ray-tracing

Object

To study the phenomena of reflection of light by a plane mirror. To study the refraction of light by plane and curved surfaces.

Background

When light traveling through air encounters a different material, e.g. a glass slab, part of the light energy is reflected back into the air and part of it is transmitted into the glass, experiencing an abrupt change in direction at the glass surface. This change in direction of the transmitted light is refraction, and the light passing into the glass is said to be refracted. We specify directions of the light rays with respect to the line drawn normal to the glass surface. Figure 1 shows an incident light ray, and the resulting reflected and refracted rays. We define the three angles as follows:

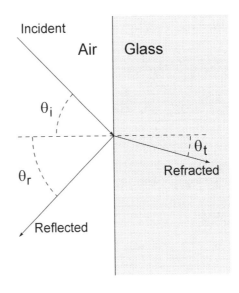

Figure 1:

- θ_i is the angle of incidence—the angle between the incident ray and the normal),
- θ_r is the angle of reflection—the angle between the reflected ray and the normal,

- θ_t is the angle of refraction—the angle between the refracted transmitted ray and the normal.

The *law of reflection* claims that the angles of incidence and reflection are the same:

$$\theta_i = \theta_r.$$

Snell's *law of refraction* asserts that the angle of incidence and angle of refraction are related through the indices of refraction of the two materials. If n_1 is the index of refraction of air and n_2 is the index of refraction of the glass slab, then

$$n_1 \sin \theta_i = n_2 \sin \theta_t.$$

The index of refraction determines the speed of light in the material, e.g. $v = c/n$, where c is light speed in vacuum. For air, we can take $n \cong 1.00$.

These two laws form the basis of geometrical optics for this lab and allow us to trace the path followed by light rays from one medium to the next.

Procedure

The three parts of this measurement should be made individually, rotating between stations, and may be performed in any order.

A. Reflection and image formation with a plane mirror. Consider light rays coming from an object, O. Each ray that falls on a mirror produces a reflected ray. These reflected rays, when extended back into the region behind the mirror, intersect at point I called the image. We wish to locate experimentally the position of the image and to check the following facts which can be proved from the law of reflection: (1) The line drawn from the object O to the image I meets the mirror at right angles. (2) The distance of the object from the mirror is the same distance from the mirror to the image.

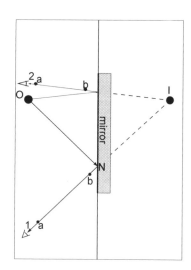

Figure 2:

Draw a line down the center of a sheet of plain paper parallel to its long edge. Use a sharp pencil. Set the front edge of the mirror on the pencil line. (The front side is the reflecting side of these mirrors.) Insert a pin at O near the edge of the page. This is the object and should remain fixed throughout the measurements. Make sure the pin is vertical.

You want to locate the image of O in the mirror. With one eye closed, look with the other at the image I in the mirror. Place two pins at points a and b so these points, your eye, and the image are all in the same line. Place one pin close to the mirror; then place another pin near the edge of the page. Keep your eye at least two feet from the mirror. When the alignment is perfect, the two pins a and b will exactly cover up the image since all are in the same line. Label each of these pin holes with the number 1. You will later draw lines representing the various incident and reflected rays and their extensions back to the image location. See Fig. 2.

In similar fashion, locate three other pairs of incident and reflected rays. To avoid confusion, remove pins a and b each time and label the associated pinholes in the paper with 2's, 3's, and 4's to identify each set of rays traced. Make sure at least two of your cases will have angles of incidence of 30° or more.

After all four ray sets have been traced, remove the mirror. Draw a line through the pin holes corresponding to b and a from ray set 1, extending it to where it intersects the plane of the mirror at point N. Draw another line from O to N. You now have a pair of incident and reflected rays. Using a protractor, construct a line normal to the mirror at point N. Measure the angles of incidence and reflection. Record these angles in a data table.

Repeat the process for the remaining three ray sets. Measure the angles of incidence and reflection and add them to the table. Do these measurements support the law of reflection?

Extend the four reflected rays $\overline{ab}$ back into the region behind the mirror (Fig. 2). See if these lines intersect at a single point. This point is the location of image I. Because of error in precisely aligning pins and images, all four lines may not intersect at the same place. If not, draw the smallest circle that includes all points of intersection between the lines. Take the center of this circle as the best average location of the image.

Measure and record the distance from mirror to image and from mirror to object. Are these distances the same? Check also if the line drawn from O to I meets the mirror at right angles.

For an alternative method of locating a mirror image, carry out the following steps. Place the mirror back on the line. Place the object at O. Remove all other pins. Look into the mirror at the image of O. Look over the top of the mirror at a second pin held in your hand behind the mirror. Place this second pin at the location of the image I. To do this, move your head from side to side; adjust the second pin so there is no relative motion between this second pin and the image. When adjustment is perfect, insert the second pin in the board. Check again by moving your head a lot from side to side. If there is no relative motion between the second pin and the image, we say there is no "parallax" and the second pin and the image are at the same place.

Mark this image location on your page and label it "Image—parallax method."

B. Refraction and Snell's law. A laser can be used to observe directly the path taken by light rays passing through different media. A block of Plexiglas will play the role a the refraction block shown in Fig. 3. A small portion of the laser light is scattered as it passes through the block, making the path evident. Outside the Plexiglas the laser beam can be traced with a pin; as the pin passes through the beam it scatters the laser light, which is then easily detected from a range of directions.

NEVER LOOK DIRECTLY INTO THE LASER BEAM OR AIM IT AT ANOTHER PERSON. LASERS PROVIDE HIGH INTENSITY COHERENT LIGHT. EVEN THE LOW POWER LASERS USED IN THIS LAB MUST BE HANDLED CAREFULLY. KEEP YOUR LASER AIMED AT THE WALL, NOT TOWARD THE CENTER OF THE ROOM OR OTHER STUDENTS' WORKING AREA.

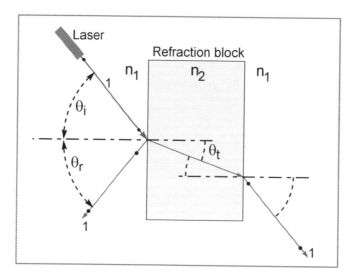

Figure 3:

Center the Plexiglas block on a sheet of paper as in Fig. 3, and draw a careful outline of the block on the paper. For three different angles of incidence, approximately 30°, 45°, and 60°, trace and mark out the path taken by the incident and refracted rays. Follow the refracted ray as it exits the block back into air. Use a pin to follow the beam in air and punch small holes in the paper at two *widely* separated points along each line segment. Take care to keep the pin vertical. Each incident ray should enter the block at a different spot to easily distinguish sets of rays. To distinguish the holes later, label each with the appropriate ray number 1, 2, or 3. Be careful not to disturb the position of the refraction block during the measurements. For the largest incident angle also trace the reflected ray from the front surface of the block.

After all the ray paths have been marked by holes, remove the block and use the holes to draw lines showing the paths taken by the laser light. For each of the three angles of incidence, draw the incident, reflected, and exiting rays and add a segment showing the path of the ray through the block.

Where each incident ray enters the block, construct a normal to the block surface. Measure the angle of incidence θ_i and the angle of the refracted transmitted ray in the block, θ_t. Do the same at the point where one of the rays exits the block. For each case use the angles to find the index of refraction of the Plexiglas block, assuming $n_a = 1.00$. Are the values of n consistent?

Verify that the one measured reflected ray satisfies the law of reflection.

C. Focal length of a convex lens.

Use the laser and a Plexiglas convex lens segment to measure the paths taken by a series of parallel rays incident from one side of the lens. Place the lens on a sheet of paper with a template for the lens and the parallel incident rays. Be sure that the lens is carefully aligned with the template. Fix the paper to the optics pin board with two pins at diagonal corners. Do not move or bump the lens once it is properly positioned.

Position the laser so the beam precisely follows dashed line #1 of the template. To better align the incoming laser beam with the lines on the template place a rectangular Plexiglas block on the template between the laser and convex lens. Make sure the face of the block that the laser beam enters is perpendicular to the template lines. Then, sighting directly down, you can make fine adjustments to the laser's position to get the laser beam correctly aligned with each line in the template in turn. Poor results will be obtained if the alignment is not done carefully.

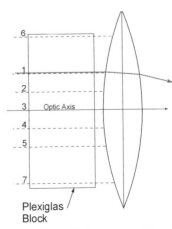

Figure 4: Parallel rays incident on convex lens.

For each incident ray use a pin to follow and mark the path of the beam after it passes through the lens. Make two pin holes in the paper. *Put one pin hole very near the lens and the second near the distant edge of the paper to create as long a line as possible.* Label each pin hole with corresponding incident ray number.

After all rays have been traced and marked, remove the lens and the paper from the board. Draw line segments using the pin holes from the lens edge to the paper's edge showing each outbound ray.

The focal length of the lens is the distance from the center of the lens to the point along the optic axis where the rays converge, the focal point. The five inner rays should all converge at nearly the same location. Measure the focal length of the lens based on the convergence point of these five rays.

The two outermost rays may cross the optic axis at a different location due to spherical aberration of the lens. How far from the focal point determined from the five inner rays do the outer rays cross the optic axis?

Before leaving lab: Leave your table neat and ready for the next lab section. Make sure each station at your table has the set of components needed for that experiment, neatly organized, and that the lasers are safely aimed toward the wall. Clean up any paper or trash from your table.

24—Converging lens

Object

To measure the focal length of a converging lens by various methods. To study how a converging lens forms a real image.

Background

The picture on the screen in a movie theater is called a "real image." It is made by light from an object passing through a converging lens as shown in Fig. 1. Light rays starting from point O on the object are bent by the lens so they all converge to meet at point I which is the real image of point O. Similarly, light from point O' converges to point I' on the image. The lens bends the rays by refraction as the light enters and leaves the lens. To produce this, a converging lens is thicker at its center than at its edge.

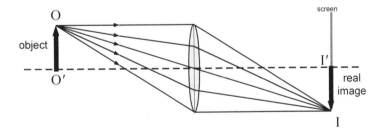

Figure 1: Some of the light rays from the tip of the object (point O) pass through the lens, which focuses them to form an image of the tip.

Note that the ray from O, which happens to fall on the center of the lens passes through it in essentially a straight line with no bending. Thus a straight line drawn from O to I passes through point C (as in Fig. 2). This fact is useful for determining the height H_I of the image in terms of the object height H_0 and the distances s and s' of object and image from the lens. Since both $\overline{OCI}$ and $\overline{O'CI'}$ are straight lines, the triangles $\overline{OO'C}$ and

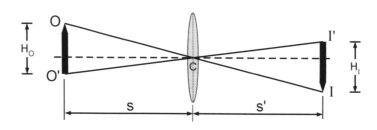

Figure 2:

$\overline{II'C}$ are similar triangles. The ratio of H_I to H_O is the same as the ratio of s' to s. The ratio of image height to object height is called the magnification M:

$$M = \frac{H_I}{H_O} = \frac{s'}{s}.\tag{1}$$

With a magnification of 8, the image is eight times larger than the object, with the image distance s' eight times s. Since the image is inverted, sometimes the magnification is reported as a negative number.

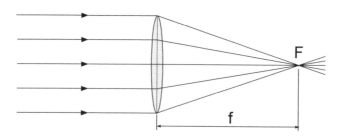

Figure 3:

In Figures 1 and 2, the dotted horizontal line is the axis of symmetry of the lens, known as the optic axis or lens axis. If we point this axis at a very distant object, such as the sun, the rays from the sun that fall on the lens are all parallel to each other and to the lens axis. These rays come to a focus at point F in Fig. 3. Point F is the "focal point" of the lens and occurs at a distance f from the lens center, known as the focal length of the lens. The focal length is a constant of each particular lens.

When the object is not extremely far away, the geometrical construction shown in Fig. 4 can be used to determine the position of the real image. To determine where light rays that originate from the tip of the object at O converge, two ray paths can be constructed.

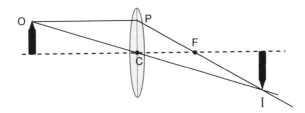

Figure 4: Geometric location of image by drawing two principal rays

(1) Draw a straight line representing a light ray from point O through point C at the lens center extending well past the lens. Rays passing through C are unrefracted so the image must lie on this line.

(2) Then draw a second ray from point O parallel to the optic axis to point P where it meets the lens. This ray must be refracted so that it passes through F. Draw the line $\overline{PF}$ and extend it beyond F. The image must also be on this line. The intersection of these two rays at point I locates the image of the tip of the object, O.

Instead of carrying out this construction every time we need to know where the image is, we can use an equation that can be derived from the construction. This is the lens equation

$$\frac{1}{s} + \frac{1}{s'} = \frac{1}{f},$$
(2)

where s = distance from object to lens, s' = distance from image to lens, and f = focal length of lens. If we wish to solve for s' in terms of the other quantities, the equation can be rewritten as

$$s' = \frac{f \cdot s}{s - f}.$$
(3)

Notice that if the object is very far away $(s \to \infty)$ the image distance reduces to f. Alternatively, the focal length can be written in terms of s and s' as

$$f = \frac{s \cdot s'}{s + s'}.$$
(4)

Procedure

Part A: Determine the focal length by various methods.

Method I-a. Have only the screen and lens mounted on the optical bench. Place the screen at the very end of the ruled scale on the base of optical bench. Put a clear light bulb—with the filament easily visible—at the opposite end of optics bench. Tweak the position so that the filament is aligned with the opposite end of the ruled scale. Adjust the position of the lens to obtain a focused image of the filament on the screen. Measure the image distance (lens center to screen) and object distance (lens to filament). Calculate the focal length of the lens.

Describe what happens to the image on the screen if you block the upper half of the lens on the incoming light side with an opaque card.

Method I-b Repeat Method I-a with a mask over the lens, allowing light through only the center half of the lens. Re-adjust the lens position to achieve a more sharply focused image. By using only the central part of the lens spherical aberration effects are reduced, but the image is dimmer since less light is used to form it. This will give a better estimate of focal length. Determine the focal length from the new image and object distances.

Method II-a. Remove the light bulb from the bench and replace it with a bright or light colored object. Keep the light bulb nearby to illuminate this object. Replace the screen with a pointer. Look over the pointer, through the lens, at the object as in Fig. 5. Keep your eye at least 30 cm from the pointer. You are looking at the back of the real image, which previously appeared on the screen. Remove the lens mask.

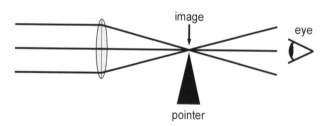

Figure 5:

You now want to put the pointer at the same position as the real image using the method of parallax. Moving your head from side to side, adjust the pointer's position until you see no relative motion between the pointer and the image of the object. When there is no relative motion, the image forms at the pointer's location.

Measure the image distance from the lens center to the pointer and the object distance. Find the focal length of the lens. This gives a better estimate of focal length.

Method II-b. We could improve on Method II-a if we could only get closer to the pointer and see more clearly if any parallax is there. To do this, hold a magnifying glass close to your eye; move closer to the pointer than in II-a, until you can see the pointer. You will also see the image because it is at the same place as the pointer (or very nearly so).

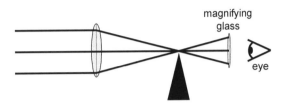

Figure 6:

Now move your head and magnifying glass from side to side and make fine adjustments to the pointer's position again until there is no discernible parallax (i.e., no relative motion) between the image and pointer. Measure the distance from lens center to pointer again.

Find the focal length and use this value as your best determination of f later in the experiment.

Method III-a. We use the fact that the image is produced at the focal point when the object is very far away (i.e. at infinity). Put the optical bench on a cart so it can be wheeled to the hallway to be aimed at objects outside. Have only the screen and lens (without mask) on the optical bench. Place the lens at a convenient cm mark. Adjust the screen to obtain a focused image on it of some object that is more than 50 meters away. For practical purposes this makes $s \gg s'$ and practically speaking $1/s = 0$ in the lens equation, so $f = s'$.

Measure the distance from lens center to screen. This is a rough estimate of the focal length.

Method III-b. Repeat Method III-a with the mask over the lens, allowing light through only the center of the lens. You will again find the focus is sharper but dimmer. This will give a better estimate of focal length than the unmasked lens, provided the image is bright enough to still see.

Compare the focal lengths found by the various methods in a table. List s, s' and f for each method in a single table summarizing your measurements.

Part B: Magnification—study of a real image.

Place the illuminated object box near one end of the bench. Set the lens at a distance between 1.4 and 1.7 times the focal length away from the object. Have the mask in place on the lens. Adjust the screen position so a focused image of the object appears on it. (See Fig. 7.)

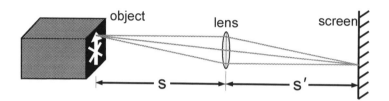

Figure 7: Magnification measurements set-up.

Measure the object distance s and the image distance s'. Measure also some dimension on the image and the corresponding dimension on the object so you can calculate the magnification $M = H_I/H_O$. Enter these dimensions in a table like the one below.

Repeat the magnification measurements with the object distance set at exactly twice the focal length. As before, adjust the screen to get a sharp focus. Measure the object and image distances. Also measure dimensions of the object and image to determine magnification.

Repeat with the object distance set at greater than twice the focal length. Adjust screen for focus. Measure image and object distances, as well as the dimensions of the object and image.

(1)	(2)	(3)	(4)	(5)	(6)	(7)	(8)
Object distance s	Image distance s'	Object height H_O	Image height H_I	Magnification H_I/H_O	Expected Mag. s'/s	Expected s' $= fs/(s-f)$	% Difference between (2) and (7)
...	...	...	...	...	...	...	...

In column 5, find H_I/H_O, which is the observed magnification. In column 6, find s'/s, which is the magnification expected by theory. In column 7 calculate image distances given by Eq. 3. Compare the predicted and experimental values by reporting percentage differences between columns 2 and 7 in column 8. Comment on the agreement of columns 5 and 6, but do not calculate percentage differences.

Graphical Method. Do the graphical construction outlined in the introduction (see Fig. 4) to find the image distance using the third case, $s > 2f$. Make the drawing fill most of a sheet of paper. *Draw an object that is actually 5 cm tall on your drawing.* Choose a suitable scale for the object and image distances along the optic axis and write this on your drawing. Draw the object at the distance used above (properly scaled) and use the ray construction to find the image distance from your drawing. Compare this distance (converted back to real units with your scale factor) to the value in column 2 of your table.

Before leaving lab: Leave your table neat and ready for the next lab section. Make sure your table has the set of components needed for that experiment, neatly organized. Clean up any paper or trash from your table.

Appendix A
Significant figures

Most of the data used in the physics laboratory are the result of measurements of physical quantities. All measurements have limits imposed by the precision of the measuring instruments and variations in the quantity being measured. In recording data or reporting the results of calculations it is important to give a consistent indication of the precision of the measurement or calculation. One way of indicating this is the correct use of significant figures. More careful treatment of experimental uncertainties and error analysis is described in another appendix.

Significant figures in decimal numbers are those digits whose values are known to be exact or of limited uncertainty. For example you could measure the length of your physics text book with a meterstick, which has 1 mm as its smallest division and get the result $L = 28.3$ cm. The digits 2 and 8 are known to be correct to a high degree of certainty. The last digit, 3, may have some error, but by reading the meterstick scale to the nearest division, there is good reason to believe that the true value is closer to 3 than it is to 2 or 4. The true value of the length of the text book is between 28.25 cm and 28.35 cm. Calibration errors of the meterstick and variations in the length of the book may cause some increase in this range, but we would still have some confidence in this third figure. It is a significant figure. The estimation and reporting of a fourth figure, such as 28.34 cm, would not be justified since the difficulty of estimating $\frac{1}{10}$ of the very small space between mm marks, the possible calibration errors in the meterstick, and the difficulty of estimating the exact edge of the book cover would make the value of this last figure very uncertain. This measurement then has 3 significant figures and would be recorded as 28.3 cm. To report the number as 28.30 cm would be incorrect as the 0 in the fourth figure would imply that figure was known with some certainty, which would not be justified in this measurement. To report the number as 28 cm would be correct to two significant figures, but would be throwing away good information and would not correctly reflect the precision of the measurement.

Zeros used to locate the decimal point in small and large numbers are not significant figures. The zeros in numbers such as 0.00462 or 364000 are not considered significant figures. When writing large numbers in the decimal form, such as the number 364000 above, there is ambiguity as to whether all of the three zeros are there just as placeholders or if one or more are significant figures. This uncertainty can be removed by writing the number in scientific notation. If all three zeros are only used to place the decimal point the number should be written as 3.64×10^5; if one zero is significant the number would be written as 3.640×10^5. If all three zeros are significant (of limited uncertainty), the number should be written as 3.64000×10^5. Occasionally large numbers may be written with a decimal point e.g., "364,000." where the decimal point is used to indicate that all digits up to it are significant. Scientific notation is considered less ambiguous and is the preferred form. In scientific notation any terminal 0's are interpreted as significant figures.

When the result of a calculation is reported, care must be taken to report the correct number of significant figures. If the calculation involves a combination of multiplication and division, the number of significant figures in the result will be the same as the datum that has the fewest significant figures. As an example, consider multiplying two measurements together: $8.431 \times 7.14 = 60.19734$. Since the datum with the fewest significant figures has 3 figures (7.14), the result should be rounded to 3 significant figures: 60.2. When a calculation requires a series of multiplication and division operations, do not round off at each step. Carry all figures on the calculator to the final operation and then round off to the correct number of significant figures; rounding off at each step throws away some useful information.

When adding a series of numbers, such as .35, and .564 with different numbers of significant figures and decimal places, write the numbers in a vertical column with the decimal points aligned and add the figures. The least (farthest right) significant figure is in the rightmost column that has no uncertain figures. Any digits beyond this are not significant in the sum.

```
  0.35
+ 0.564
-------
  0.914
```

In this example the first position with an uncertain value would be the third place to the right of the decimal point. The correct two significant figure result would be 0.91. The last digit was dropped since it was uncertain in one of the numbers.

As a second example add the numbers 1.7, 2.44, and 2.23.

```
  1.7
  2.44
+ 2.23
------
  6.37
```

This time we again have a two significant figure result but since the third figure is greater than 5 we round the figure up to get our correct two significant figure result of 6.4.

Do not drop the figures that exceed the minimum significant figure position or round the numbers to the minimum number of significant figures before addition since this may result in throwing away some information. In the last example that would result in a two figure number $1.7 + 2.4 + 2.2 = 6.3$, which would be throwing away the information that the last two given numbers add to closer to 6.4 than to 6.3. Subtraction follows the same rules.

For your laboratory work, the ultimate test of whether a digit in your measurement or result is significant can be accomplished by asking yourself the question, "If I (or another person, using the same equipment and equal care in the measurements) repeat the experiment, will I arrive at the same result?" If your result is $R = 3.567$ in a first trial and $R = 3.425$ in a second trial, you cannot claim to have four significant figures in your result, regardless of how many significant figures you think the original measurements had. At best you might claim that $R = 3.5 \pm 0.1$. Two significant figures are viable, since the measurements suggest that the second figure, while not known with complete confidence, appears to have limited u̲ncertainty. Any more digits in the result are unwarranted.

More thorough methods for estimating uncertainties in results are described in a later appendix and will be required in some of the experiments you perform.

Another point to remember is that you generally cannot have a result with more significant figures than any individual measurement contributing to the result. For example, if you measure mass to four significant figures, $231.3\,\text{g}$, and two temperatures to three significant figures, $T_i = 22.5°\text{C}$ and $T_f = 18.3°\text{C}$, and the result depends on the product of the mass and the temperature difference, $T_i - T_f$, you cannot claim more than *two* significant figures. The temperature difference is known at best to two figures: $4.2°\text{C}$. This limitation then controls the precision of your result.

Then $R = m\Delta T = 231.3\,\text{g} \times (22.5 - 18.3)\,°\text{C} = 97\,\text{g}°\text{C}$ – not 97.146, or even 97.1. In fact, if $\Delta T = 4.1\,°\text{C}$ on a second trial, the new result would be $95\,\text{g}°\text{C}$. A change in the last measurable digit of ΔT (tenths of a degree) produces a change in the 2nd digit of the result. The result has at best two significant figures.

Appendix B
Error analysis

Laboratory instruments and the experimenter's skills have limits when making measurements. Each measurement can be characterized by some uncertainty or probable error. When reporting an uncertainty or probable error, the experimenter states a range of values in which he or she expects repeated measurements will fall. Any result calculated from the measurements is also characterized by some uncertainty. A question that naturally arises in the lab is "How does the uncertainty in the measurements made contribute to the uncertainty in the final result calculated from the measurements?"

As an example, a block of wood is measured to find its length (L), width, (W), and depth, D. It is weighed to find its mass, M. Each of these measurements has some uncertainty. If the density is calculated from the measurements $\rho = \frac{M}{LWD}$, how does the uncertainty in the density depend on the uncertainties in the individual measured quantities?

This type of error analysis is different from simplistic error calculations you almost certainly have encountered in laboratory courses where a 'percentage error' is calculated based on a comparison of your result with an accepted "true" value. While this sort of "error analysis" is a useful tool in judging how carefully you have done a controlled pedagogical exercise, in real laboratory work you generally don't know the "true" answer. If you did, why would you bother with the measurement? Instead, you want to be able to tell others how confident you are of your answer. The kind of error analysis discussed here is focussed on assessing the uncertainties left by your measurements.

Making realistic estimates of uncertainties in measurements and applying the rules for how these uncertainties propagate to the final result are important laboratory skills. Some simple techniques are described here.

Types of errors

The errors that occur in making measurements in the laboratory can be categorized into three general types.

- <u>Gross errors or blunders</u> are due to misreading, using wrong scales, incorrect procedures, etc. The results generally cannot be salvaged. These errors require you to repeat an experiment, probably after designing improved procedures.

- <u>Systematic errors</u> might be corrected for after the fact—or avoided entirely by more careful procedures. A review after the experiment of instrument limitations or a more complete analysis of the data collected may allow you to make reasonable corrections. Examples of this type of error are instrument mis-calibration, changes due to drifts in the ambient temperature, parallax in reading scales. Systematic errors often shift a result in a consistent and predictable direction. Unfortunately, it isn't always obvious that an experiment includes a systematic error and it may go undetected!

- <u>Random errors</u> are due to effects that are frequently beyond the control of the experimenter. They are equally likely to add to or subtract from the measurement.

Error intervals

An important part of any measurement made in the laboratory is the estimation of the precision of the measurement. This is usually indicated by the use of the uncertainty or "error interval." The error interval is an estimate of the uncertainty or probable error, which when added and subtracted from the best value for the measurement, gives the range of values in which the experimenter expects the true value must fall. The results of a measurement would be reported in the form $x \pm \Delta x$, where Δx is the best estimate of the error interval or probable error.

Estimating errors

The estimation of the likely error interval for systematic errors involves the careful analysis of the apparatus and the procedures used in the measurement. This requires knowledge of the calibration of the apparatus, scale divisions, sensitivity, etc. Experience plays a large part in the estimation of a suitable error interval. Beginners in laboratory work are inclined to underestimate the size of the error interval due to systematic errors. The size of the error interval should be chosen large enough that the experimenter is fairly confident that a more accurate measurement will fall within the interval, but small enough that it reflects the true accuracy of the actual measurement made

A simple rule that can be used to estimate the error in a measurement made only once, or has the same value on repeated measurements, is to take $\frac{1}{2}$ the smallest division that can be read on the scale. For example if a coin's diameter is measured with a ruler whose smallest division is 1 mm, the diameter might be reported as $d = 12. \pm 0.5$ mm. This assumes there is no calibration error in the ruler and expresses the experimenter's judgment that the true diameter is closer to 12. mm than to 11. mm or 13. mm. Sometimes it is reasonable to attempt to make a finer reading between the smallest subdivisionsand report a smaller uncertasinty. So it might be possible to instead say $d = 12.3 \pm .03$ mm. The conservative approach is to settle for half the smallest division.

Just because an instrument provides many digits in its measurement is not a guarantee that they are all significant or meaningful digits. Sometimes there is a stated accuracy of the calibration of an instrument or of standards such as masses. If this uncertainty is larger than $\frac{1}{2}$ the smallest division or other estimate of error (such as statistical estimates as discussed in the next section), then this would be the better estimate.

For example, a balance has smallest divisions of 0.01 g. You find $m = 240.06$ g when a mass is weighed. It would be tempting to report the mass and uncertainty as $m = 240.06 \pm 0.005$ g. But, if the manufacturer also reports that the accuracy of the balance is not guaranteed beyond 0.05% of the reading, the correct estimate of the uncertainty is 0.05% of 240.06 g or 0.12 g! The correct way to report the measurement would then be $m = 240.06 \pm 0.12$ g This means that if you were to give the mass to someone else with the same or better model balance, you are reasonably confident that they will find $239.9 \leq m \leq 240.2$ g. While reporting 5 digits in a reading, this balance would probably be good to only 4 figures for this measurement.

Estimating random errors

When repeated measurements of the same quantity give different values and the variations in the measurements are due to randomly varying sources of error, statistical methods are used estimate an error interval. Only a brief summary of important concepts follows. For most purposes the best value of the measured quantity in the presence of randomly varying measurements is the mean or average. The mean of x, denoted by $\bar{x}$, is defined for a set of n individual measurements x_i as

$$\bar{x} = \frac{1}{n}\sum_{i=1}^{n} x_i = \frac{x_1 + x_2 + \cdots + x_n}{n}. \tag{1}$$

The deviation of an individual measurement is the the difference between it and the mean of the entire set of measurements: $x_i - \bar{x}$. A very simple measure of the uncertainty or probable error in a set of measurements can be found from the mean of the absolute value of the deviations:

$$\overline{\Delta x} = \frac{1}{n} \sum_i |x_i - \bar{x}|. \tag{2}$$

A more sophisticated measure of the probable error is the standard deviation, σ. σ is defined by

$$\sigma_x^2 = \frac{1}{n-1} \sum_{i=1}^n (x_i - \bar{x})^2. \tag{3}$$

The quantity σ^2 is usually called the variance. The standard deviation of a set of measurements is very nearly the square root of the average of the squared deviations. (We use a definition where we divide by $n-1$ instead of by n.) Most scientific calculators can calculate the standard deviation for a set of measurements as a built-in function. Refer to your calculator's manual if you wish to make use of this feature.[1]

If you calculator does so, you can use it to estimate σ. The simpler method of estimating the statistical error by using the mean deviation can be used when statistical functions are unavailable in a calculator. An example of finding the mean, mean (absolute) deviation, and standard deviation for a set of length measurements, in cm, is shown below.

Measurement #	Result L_i (cm)	$|$Deviation$|$ $|L_i - \bar{L}|$ (cm)	Deviation sqr'd. $(L_i - \bar{L})^2$ (cm^2)
1	5.16	0.026	6.76×10^{-4}
2	5.13	0.004	0.16×10^{-4}
3	5.15	0.016	2.56×10^{-4}
4	5.12	0.014	1.96×10^{-4}
5	5.11	0 .024	5.76×10^{-4}
Mean	5.134	0.024	$\sigma^2 = 4.3 \times 10^{-4}\,\mathrm{cm}^2$
Rounded	5.13	0.02	$\sigma = 0.02\,\mathrm{cm}$

---

[1]Some calculators make a distinction between the variance σ^2 and sample variance s^2. s provides a more pragmatic estimate of the probable errors in measurements and is consistent with the definition used here for σ. The difference between these two vanishes as the number of measurements n becomes large. s is what you get from a finite number of measurements, σ is what you *would* get if you had time to make an infinite number of measurements. Luckily for you, physics labs only *seem* like an eternity; you will always in practice be calculating and using s as an estimate of the unattainably ideal σ. In the unlikely event you are reading this appendix, have pursued this footnote, and *still* have some curiosity left over, you are almost certainly a genuine geek. See a statistics text for details on these subtleties. The short version is that this definition makes σ undefined if there is only one measurement, $n = 1$; with one measurement you have no basis for estimating random errors and have no business reporting an uncertainty based on statistical arguments.

The value of the measurement and the estimated error would be stated as $L = 5.13 \pm .02$ cm. Here $\sigma = 0.0207$ before rounding. The mean deviation usually over-estimates the uncertainty compared to σ.

A useful quantity in characterizing a measurement is the *fractional* error or uncertainty. Let's define the fractional uncertainty in x in terms of the mean value $\bar{x}$ and the uncertainty Δx, which may have been found by calculating the standard deviation, σ_x, the mean deviation, or by some other estimate. The fractional error is

$$e_x = \frac{\Delta x}{|\bar{x}|}. \qquad (4)$$

The percentage error is $100\% \times e_x$. So, if $\bar{x} = 2.02$ and $\Delta x = 0.06$, the fractional error in x is $\frac{.06}{2.02} = 0.03$, and the percentage error or uncertainty is 3%. The fractional error turns out to be useful when combining various sources of error, as you will see below.

Propagation of errors

The determination of the value of a physical quantity such as density, acceleration of gravity, etc., which is the usual objective of a laboratory experiment, is seldom obtained by one direct measurement. The desired quantity is usually related in some known way (an equation) to two or more measurable quantities and is calculated from the measured values.

This brings us back to the central question: "How does one estimate the error in the calculated quantity from the estimated errors in the directly measured quantities?" In other words, how do the errors in the measured quantities "propagate" through the known relation to produce an error in the calculated quantity? If the relationship is given in the form of a mathematical equation then the methods of differential calculus can be used to find the probable error in the calculated quantity. But we'll begin with a simple rule that overestimates the probable error.

"Crank-three-times" rule. This is a very simple way to evaluate the worst case scenarios. Consider determining the density, ρ, of a rectangular block with sets of measurements for the mass, M, and the block's three dimensions, length L, width W, and depth D, so you can calculate averages for each of these quantities. The *best* estimate for the density ρ is calculated from the average values:

$$\rho_{best} = \frac{\bar{M}}{\bar{L}\bar{W}\bar{D}}.$$

The *smallest* possible density from these measurements would be calculated using the smallest individual mass measurement and the largest of each of the dimension measurements:

$$\rho_{min} = \frac{M_{min}}{L_{max}W_{max}D_{max}} .$$

And the *maximum* possible density would follow from using the largest individual M measurement and the smallest from each of L, W, and D:

$$\rho_{max} = \frac{M_{max}}{L_{min}W_{min}D_{min}} .$$

So this amounts to plugging numbers into the density formula $\rho = M/V$ and turning the calculational crank for three cases, hence the name "crank-three-times." This may substantially over-estimate the probable error, but is relatively easy to apply in cases where the better methods become cumbersome.

Improved combination of uncertainties. Take the two-variable case where the quantity z is a function of two measured variables x and y: $z = f(x, y)$. Let's assume measurements of x and y have yielded $\bar{x}$ and $\bar{y}$ and uncertainties Δx and Δy. Taking the differential of the function gives[2]

$$dz = \frac{\partial f}{\partial x} dx + \frac{\partial f}{\partial y} dy.$$

The best estimate of $z = f(\bar{x}, \bar{y})$. The largest change in z from this best value caused by the uncertainties in x and y would be

$$\Delta z = \left| \left(\frac{\partial f}{\partial x} \right)_{\bar{x}, \bar{y}} \right| \Delta x + \left| \left(\frac{\partial f}{\partial y} \right)_{\bar{x}, \bar{y}} \right| \Delta y. \tag{5}$$

The partial derivatives are evaluated at $x = \bar{x}$ and $y = \bar{y}$, as denoted by the subscripts. This provides the estimate of how much uncertainty there is in z from the uncertainties in x and y. The absolute values of the partial derivatives ensure that our estimate of the uncertainty Δz is always positive.

If the function f is addition, so $z = x + y$, then the necessary partial derivatives are just constants, $\frac{\partial f}{\partial x} = 1$, $\frac{\partial f}{\partial y} = 1$, and the uncertainties add directly:

---

[2]The *partial* derivative $\frac{\partial f}{\partial x}$ is found by differentiating f with respect to x, $(\frac{df}{dx})$ while treating y just like a constant. Similarly, $\frac{\partial f}{\partial y}$ is the derivative of f with respect to y, $(\frac{df}{dy})$, while treating x like a constant.

$$z = x + y \quad \Longrightarrow \quad \Delta z = \Delta x + \Delta y.$$

If the function is a product, $z = xy$, then $\frac{\partial f}{\partial x} = y$, and $\frac{\partial f}{\partial y} = x$, so the uncertainty in z calculated from measurements of x and y is

$$\Delta z = |y|\Delta x + |x|\Delta y$$

Dividing by the function $|z| = |xy|$ puts this in terms of fractional errors:

$$z = xy \quad \Longrightarrow \quad \frac{\Delta z}{|z|} = \frac{\Delta x}{|x|} + \frac{\Delta y}{|y|} = e_x + e_y$$

When two quantities are multiplied, their fractional (or percentage) errors *add*. This is a handy rule to keep in mind. The values of x and y used to evaluate this expression would again be $\bar{x}$ and $\bar{y}$ if based on repeated measurements.

For a quotient, $z = \frac{x}{y} = xy^{-1}$, the product rule applies for computing fractional errors, even though one partial derivative introduces a minus sign. Independent errors in general cannot be expected to cancel out. Instead, the absolute values of uncertainties or fractional errors must be added.

$$z = \frac{x}{y} \quad \Longrightarrow \quad \frac{\Delta z}{z} = \frac{\Delta x}{|x|} + \frac{\Delta y}{|y|} = e_x + e_y$$

For a power function, $z = x^n$, $\Delta z = nx^{n-1}\Delta x$. Dividing by the function, x^n, puts the result in terms of fractional errors again.

$$z = x^n \quad \Longrightarrow \quad \frac{\Delta z}{z} = |n|\frac{\Delta x}{|x|}.$$

These results have straight-forward extensions to functions of more than two variables and can be applied to complicated equations by applying them in succession. As an example consider the equation used to calculate the density of a metal cylinder. The cylinder has a length L, diameter d, and mass M. The volume of a cylinder is the circular area times the length: $V = \pi \left(\frac{d}{2}\right)^2 L$. The density is

$$\rho = \frac{M}{V} = \frac{4M}{\pi d^2 L}.$$

The factors M, d^2, and L are all handled with the product or quotient rules. The power rule is applied to the d^2 term, leading to the equation

$$\frac{\Delta\rho}{\rho} = \left|\frac{\Delta M}{M}\right| + \left|\frac{\Delta V}{V}\right| = \left|\frac{\Delta M}{M}\right| + 2\left|\frac{\Delta d}{d}\right| + \left|\frac{\Delta L}{L}\right|.$$

ΔM, Δd, and ΔL would be the estimates of the uncertainty obtained either from standard deviations, or from resolution limits imposed by the measuring device, or other appropriate estimates.

In most experiments the assumption is made that the probable errors in the various measured quantities are all added so that they maximize the total error. This scenario is usually too pessimistic, leading to an over-estimate of the probable error of the calculated quantity. However, for the purposes of these labs, this technique will be the usual procedure because of its simplicity. If you want fancier techniques, read on, but they won't be needed for this course.

More detailed treatment of random error propagation (optional)

If the errors in the measured quantities are random, then there *is* on average going to be some cancellation of errors from different quantities since there is an equal probability of an average measured value being above or below the true value. A brief discussion of improved methods for combining random errors follows.

In the case of *random errors*, where the error estimates are the standard deviations, the better equation for adding together the various errors in $z = f(xy)$ is

$$\sigma_z^2 = \left(\frac{\partial f}{\partial x}\sigma_x\right)^2 + \left(\frac{\partial f}{\partial y}\sigma_y\right)^2.$$

This can be applied to the various common combinations treated earlier.

- For sums or differences:

$$z = x \pm y \longrightarrow \sigma_z^2 = \sigma_x^2 + \sigma_y^2.$$

  (Notice that the uncertainties add like the perpendicular components of a vector to form the magnitude of the resultant.)

- For products or quotients:

$$z = xy \text{ or } z = x/y \longrightarrow \left(\frac{\sigma_z}{z}\right)^2 = \left(\frac{\sigma_x}{x}\right)^2 + \left(\frac{\sigma_y}{y}\right)^2 = e_x^2 + e_y^2.$$

  (Here the *fractional* errors add like the perpendicular components of a vector.)

- For power laws:

$$z = x^n \longrightarrow \left(\frac{\sigma_z}{z}\right)^2 = \left(n\frac{\sigma_x}{x}\right)^2.$$

Applying these new rules to the example of the calculation of density results in the following:

$$\frac{\sigma_\rho}{\rho} = \sqrt{\left(\frac{\sigma_M}{M}\right)^2 + \left(2\frac{\sigma_d}{d}\right)^2 + \left(\frac{\sigma_L}{L}\right)^2}.$$

This more careful statistical treatment of error propagation gives smaller estimates of the uncertainty than the simpler worst-case approach. For the density calculation example, if each of the measured quantities had an error of 1% (e.g. $\sigma_M/M = 0.01$, etc.), the worst-case approach gives

$$\frac{\sigma_\rho}{\rho} = 0.01 + 2 \times 0.01 + 0.01 = 0.04 \quad (4\%).$$

The new improved propagation rule gives instead

$$\frac{\sigma_\rho}{\rho} = \sqrt{(0.01)^2 + (2 \times 0.01)^2 + (0.01)^2} = 0.024 \quad (2.4\%),$$

which is only slightly larger than the contribution from the largest term in the expression. The result of adding as orthogonal components is frequently to make the largest error term dominate the expression. If one error term is much larger than the other terms, the smaller terms can be neglected, which results in a much simpler expression.

These equations can be used for propagation-of-error calculations even if the error estimates are not the standard deviations obtained from multiple measurements.

Appendix C
Graphing data

Presentation of experimental data in the form of graphs is perhaps the most efficient method of presenting the results of an experiment. It is far easier to recognize possible correlations between the two measured quantities when displayed on a graph than when displayed in tabular form as simply two (or more) columns of numerical values. A graph instantly reveals any qualitative trends that may link the "dependent" variable (usually the quantity that specifies the y-coordinate or ordinate of a data point) and the "independent" variable, (usually plotted along the x-axis or abscissa). Carefully drawn graphs of data will also facilitate quantitative comparisons of the experimental data with various models or theories that seek to explain the data.

Graphs may be prepared in a variety of ways and in a variety of formats. Hand-made plots on graph paper are entirely satisfactory when carefully done. In situations for which the number of data points is not large, plotting by hand is efficient and provides a better feel for the data by the experimenter. For large quantities of data it is convenient to generate the plots with the aid of software. Regardless of the means used to produce the plot, an effective and appropriate graph has a number of important features. If you use software, make sure you know how to customize the graph's properties and that you understand how to produce the desired plot.

Essential features

- *The plotted data should fill as much of the page as possible.* Choose the range of values along each axis large enough to accommodate all the data, yet not so large as to result in the data occupying only a corner or other small portion of the graph. Select ranges for the two axes so that the experimental data will occupy as much of the graph paper or page as possible.

- *Include clearly drawn and labeled axes.* These labels consist of two parts. First is a clear indication of the range of values allowed along

197

each axis with numerical values at uniform intervals denoted by "tic-marks." Second is a title along each axis specifying what physical quantity is represented by that axis. If the axes are not labeled, there is no way for the reader to easily determine what the graph is supposed to represent. Axes should be labeled and the units of the measurement specified on the plot itself. There should be no doubt in the reader's mind about where the axes on the plot are and what they represent.

- *Choose a convenient interval between tic-marks that facilitates accurate plotting of the data and easy determination of the coordinates of plotted points by the reader.* On pre-printed graph paper, choose tic-mark intervals that make quick calculations easy using the printed grid. If using software, force the software to choose convenient numbers for the ranges of the axes and for the tic-mark intervals. Don't settle for poor or irregular default values the software may provide.

- *The data points should be represented by clear, neatly drawn symbols.* The symbols should not be overly large. In the case of a graph in which several sets of data are presented simultaneously, each distinct set of measurements should be represented by a separate symbol so as to identify which data are related as a set. For data sets in which the density is such that any reasonably sized symbols will overlap, it may be sufficient to simply plot points (or small filled circles).

- *Put a title at the top of the graph to identify what it is.* Make this truly descriptive, not some generic or vague title.

- ***Don't*** *draw a jagged line connecting point to point.* Smooth lines through the data points should appear only when they serve a particular purpose. For example, such a line may be a straight line drawn through the data to represent the best fit to a linear equation modeling the data. Or a smooth curve may be drawn through the data (not necessarily through each and every data point), to serve as a "guide to the eye" to emphasize a trend in the data or to make multiple sets of data plotted on the same graph more easily distinguished. A smooth continuous curve may also be plotted to represent the results predicted by some particular model or theory for the phenomenon under study for easy comparison to the meaured data.

- *Include error bars when appropriate.* Experimental quantities are subject to uncertainties as a consequence of the finite resolution afforded by the measuring technique. The presentation of the data should indicate the uncertainties in the data being plotted. In some circumstances, the size of the symbol drawn is already larger than the estimated uncertainties and the addition of error bars is unnecessary. However, when the uncertainties are large (i.e., a significant fraction of the quantities being plotted), the error bars should be indicated. Note that error bars may be needed along both the vertical and horizontal axes of an x-y plot. Sometimes it is sufficient to indicate the typical uncertainties by affixing error bars to one data point to indicate the approximate uncertainty that applies to all the data.

- *Provide a legend for different symbols.* Whenever a plot contains multiple sets of data and therefore different symbols, or various lines for theoretical models, a key or legend should be presented either on the graph or in a nearby caption, identifying each important item.

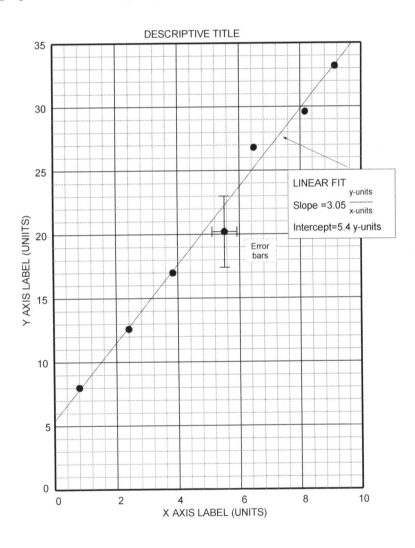

Figure 1: Typical layout of essential elements for graphs.

Graphical analysis

Plotting data may require some analysis or manipulation of the raw measurements to produce the most useful plot. For example, consider the data acquired in measuring the position of a falling body as a function of time (a set of t and y values). In introductory physics labs, such data may be obtained as a series of points along a waxed paper tape generated by a "spark timer" that marks the position of a falling body at regularly spaced time intervals. Assume for convenience that y is measured positively in the downwards direction, and that the origins of y and t are taken to be the first such mark on the tape, so the marks on the tape do reasonably represent position as a function of time. In general the first data point on the tape (defined to be $(t, y) = (0, 0)$ by choice) need not correspond to zero velocity, since the object may have begun its descent before the first data point was recorded. A simple model (constant acceleration) predicts that the position should be given by a simple expression, $y = y_o + v_o t + \frac{1}{2} g t^2$, where y_o is the initial position at $t = 0$ (here defined to be 0), v_o is the initial speed, and g is the acceleration due to gravity.

A plot of the data might look like Fig. 2. It reveals some curvature and therefore indicates some acceleration is taking place but does not provide easily interpreted evidence of the constant acceleration employed by the model. The data may be analyzed and plotted in a slightly different fashion, however. Given the (t, y) data it is straightforward to produce instead a table of values of $(t, y/t)$ and plot them, as in Fig. 3. Such a graph clearly reveals a reasonably good linear relationship between the experimentally obtained y/t and t. Furthermore, by taking the prediction of the simple model above and dividing both sides of the equation by t, one finds that the model predicts simply $\frac{y}{t} = v_o + \frac{1}{2} g t$. (Recall that y_o has been taken to be zero by virtue of how the origin for y was chosen.) The quantity y/t is sometimes referred to as a compound variable. Hence the model predicts that a plot of y/t vs. t should be a straight line with a (y/t)-intercept of v_o and a slope of $g/2$. Note that it is possible to read off from the graph directly the initial speed, and by measuring the slope of the line, to obtain a value for the acceleration of gravity, g. A simple manipulation of the raw data, when plotted, has provided very visible and quantitative verification of the prediction of the model - verification not available from the plot of the original raw data. In extracting the slope in the example above in order to obtain a value for the acceleration of gravity, it is important to keep proper track of units in order to obtain physically meaningful and accurate results.

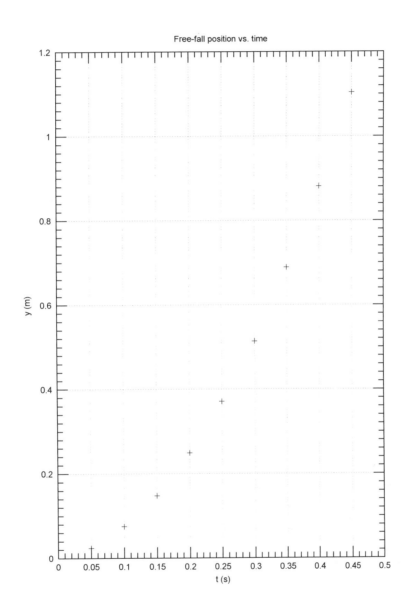

Figure 2: Example graph of free-fall data.

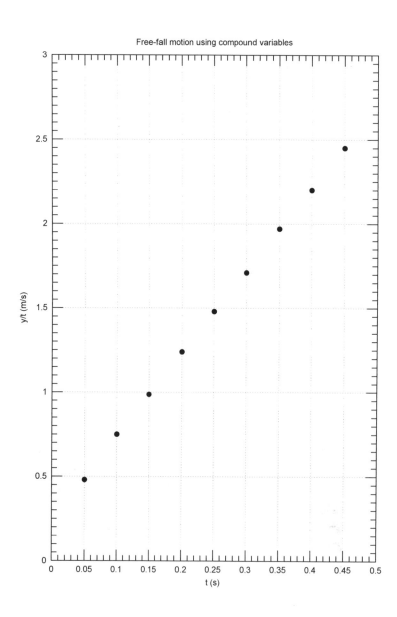

Figure 3: Plot of data using linearized free-fall model variables.

Finding slopes

Once the data has been cast into a linear form, perhaps using compound variables (X, Y), the parameters that describe the straight line $Y = mX + b$ can be found from measurements on the graph: the slope m, and the Y-intercept b. If the data appear to fall along a straight line, draw with a ruler the line that best fits the data, passing among the data points. If there is substantial scatter in the data, draw a line so that the points are about evenly distributed above and below the line all along the line's length. Extend the line to the boundaries of the graph. To find the slope, pick two points on the line *that are not any of your measured data points*. Read off the coordinates (X_A, Y_A) and (X_B, Y_B) of these two points. Then calculate changes $\Delta Y = Y_B - Y_A$ and $\Delta X = X_B - X_A$ between these two points and find the slope of the line $\frac{\Delta Y}{\Delta X}$. Include the units! If the vertical axis corresponds to $X = 0$, you can read the Y-intercept b directly from the graph. If the vertical line corresponding to $X = 0$ doesn't fall within the plotted range, you will need to extrapolate the line to find b.

Logarithmic scales

Such plots discussed above correspond to "linear-linear" plots—both axes are scaled linearly with the data. In some circumstances it may be more useful to use a logarithmic scale for one or both of the axes. Log scales are useful when the range of data is very large, spanning several orders of magnitude (or decades).

Semi-log plots are also useful in situations in which exponential processes are expected to occur. For example consider the level of radioactivity exhibited by a short-lived isotope. The measured activity may come from a Geiger counter that perhaps counts beta particles (electrons) that are emitted during the decay of the isotope under study. The number of decays per second may be expected to follow an exponential decay in a simple model, so that the reading of the Geiger counter would be predicted to obey $N_d(t) = Ce^{-t/\tau}$, where N_d is the number of decays detected per second by the Geiger counter, as read from the counter. The constant C will depend upon how much radioactive stuff is present at the start, and τ, the decay time constant, is a measure of how fast it decays; small values reflect rapid decay. This predicts that the number of decays per second will decrease exponentially with time as the measurements are carried out. If N_d is measured over time and a plot of N_d vs. t is drawn up, the data should exhibit a downward curving trend. A plot of $\ln N_d$ vs. t should be a straight line with a slope of $-1/\tau$: (If instead one takes the $\log_{10}(N_d)$ the slope will be $-\frac{\log(e)}{\tau}$.)

When faced with data that span several orders of magnitude along each axis it may be necessary to resort to a log-log plot. This is a graph in which both axes are scaled logarithmically. This technique is also useful when testing for general power-law relationships between the measured quantities—i.e., when the measured (x, y) values are thought to be related by a law of the form $y = ax^b$, where the value of the exponent b may be predicted by the model under test. The preparation of log-log plots is similar to semi-log plots. One can take the logs of both sets of data (x and y) and plot these numbers on regular graph paper, or one can use log-log graph paper and plot the original (x, y) data, or even better, use a computer to help in in the whole process of graphing. Good software will allow either of the manual approaches to be implemented as desired.

A few additional comments regarding log axes are worthwhile. The first point is to recognize that the log is defined for dimensionless quantities. For example, the absolute temperature, T, is measured in K (kelvins) and the quantity log(kelvin) has no meaning. By writing $\log(T)$, one usually means that the log is taken of the numerical value of the temperature when that temperature is expressed in units of K. This is best made explicit in labeling an axis, e.g. instead of simply $\log(T)$, writing $\log(T(K))$, to make it clear to the reader how the axis is generated. Equivalently one can think of the process as dividing the measured data by the unit of measure so, for example, that by $\log(T)$ one really means $\log \frac{T}{1K}$, when T is measured in kelvins. Often it is obvious from the context what the "normalization" is, but it usually doesn't hurt to add the unit of measure for clarity.

A second point is in regard to extracting slopes from semi-log or log-log plots to determine values for parameters used by a particular model (e.g. τ or b used above). Often the slope to be determined may be the slope of a straight line drawn simply by eye to best represent the data. To find the slopes on data plotted by directly computing logs before plotting and that data plotted using linear axes, the process is straightforward and can proceed analogously to that employed for a linear-linear plot. For plots employing log-log or semi-log axes offered by many softwatre packages or (found on semi-log or log-log graph paper) one must be more careful. When using this method, think and proceed carefully to get the proper conversion factors, or, to be safe, manually calculate logs as needed and calculate slopes as for a linear plot. For log-log plots on log-log paper it may also be possible, with a little bit of thought, to read off the exponent by inspection. If software is used for graphing purposes, often the software will allow a "least squares fit" of the data to a particular model and produce automatically the values of the parameters that provide the best agreement with the measured data. These are very convenient routines but should not be used blindly.

Appendix D
Least squares methods

In analyzing data from experiments, one often seeks a way of plotting the data so that the data appear to fall along a straight line. It is a very rare experiment when all the data fall precisely on a line; uncertainties in the measurements and errors in the measuring technique will usually contribute some scatter to the data. Usually the quantities plotted in the graph (i.e., some measured X and Y values) are selected based on some physical model. If a graph appears to form a straight line, one can conclude that a model that relates Y to X can be expressed as:

$$Y = mX + b \tag{1}$$

This is, of course, just the general form for the equation of a straight line; m is the slope of the line and b is the Y-intercept. The physical model that leads to this equation usually provides an interpretation of the physical significance of the slope and intercept. For example, consider an object falling freely under the influence of gravity. Its displacement can be expressed as a function of time by the familiar expression:

$$y - y_o = -\frac{1}{2}gt^2 + v_o t \tag{2}$$

where g is the acceleration due to gravity, v_o is the initial velocity, and y_o is the initial position. In a typical experiment, one would measure and record a series of values for $y - y_o$ and t. The physical model represented by eq. (2) can be massaged into a linear form by dividing both sides of the equation by t:

$$\frac{y - y_o}{t} = -\frac{1}{2}gt + v_o \tag{3}$$

and then defining new variables: $X = t$ and $Y = \frac{y-y_o}{t}$ (X and Y are sometimes called compound variables) so that the "linearized" equation can be written as:

$$Y = -\frac{1}{2}gX + v_o \tag{4}$$

This now looks like the equation of a straight line. If one takes free-fall data, i.e., a set of raw, measured y_i and t_i values, computes from these the corresponding X_i and Y_i values, and then plots the resulting Y_i as a function of X_i, the graph should resemble a straight line if the model (eq. (2)) is applicable to the experiment. By direct comparison of eq. (4) with eq. (1) above, it's simple to recognize that the slope of the resulting straight line must correspond to $-\frac{g}{2}$, and the Y-intercept must provide a direct measure of the initial velocity, v_o. So by measuring the slope from the graph and by reading off the intercept, the parameters (g and v_o) of the physical model (constant acceleration motion) can be determined. The question next arises of how best to determine the slope and the intercept of the plotted data. Again, most experiments will show some scatter in the plotted data, and it will not usually be obvious exactly what line to draw, since a single line is unlikely to pass through all the data. How does one decide what line to draw in order to measure the slope and intercept from the graph? This question can be addressed in two different ways. The first approach is a qualitative method that is quite simple and quick to implement, but lacks rigor.

(1) Using a ruler, draw a straight line that best fits the data or most nearly represents the largest number of data as judged by eye. This line need not pass through all the points, (in fact the best line may not pass directly through any single datum) but the points, if not falling on the line, should be distributed nearly equally above and below the line along the entire length of the line. Since there is limited precision to the measurements, it is not surprising that all the data don't sit exactly on a line. Once the line is drawn, using the line and *not* any actual data points, measure the slope and Y-intercept of the line. This method requires that careful attention be paid to the units of the the the quantities (X and Y) involved. It is also important to use the drawn line in calculating the slope and not a particular pair of data points, since using specific data would defeat the purpose of drawing the line, which is to reflect in some sense the average slope of the data.

(2) A more rigorous method is to introduce a quantitative measure of how well a particular line imitates the measured data. In the ideal experiment, the measured data and the straight line would coincide exactly. The distance of each measured datum from the ideal line would be zero. Because of the scatter present in real world data, the measured data points will fall above and below the "best" line. As a result, each measured datum is some distance from the line. If one adds up the distance each measured point is from the straight line, one has a measure of how well the line represents the data. The better the fit, the smaller the "total distance" between data and line. It might seem that one could spend an eternity trying different lines

with slightly different slopes and intercepts in a quest to find the ultimate best-fit line. Fortunately, there is a well-defined procedure for determing the optimum slope and intercept to represent the data.

This systematic method is the least squares method. Assume a set of data has been collected (such as the free-fall example described earlier) and a set of compound variables, X and Y, have been found based on a physical model, such that the model, when written in terms of X and Y, is in the form of the equation of a straight line ($Y = mX + b$). Then it is possible to take each of the values of X_i from the measured data and compute a value of Y for given values for m and b. Call these calculated values Y_{calc}. $Y_{calc}(X_i) = mX_i + b$ is then the value of Y predicted using the model with specific choices for m and b and $X = X_i$. The least squares process is a means of finding the values of slope and Y-intercept (m and b) that minimize the quantity:

$$\chi^2 = \sum_{i=1}^{N} \left(Y_{calc}(X_i) - Y_i \right)^2 \tag{5}$$

This quantity is a direct measure of the (vertical) distance of all the data points from the straight line with slope m and intercept b. $Y_{calc}(X_i) - Y_i$ measures the difference between the model's predicted line and the actual measurement. This quantity is squared to make irrelevant whether a particular measured datum falls above or below the predicted line; its *distance* from the calculated line is the important quantity. N is the number of data points and the Y_i are the measured values from the experiment. Note that when χ^2 (chi-squared) is zero, all the calculated values and measured data match exactly; the calculated line passes exactly through each measured point.

How can χ^2 be found if the optimum m and b are not yet known? The best values of m and b can be found by minimizing χ^2 with respect to both m and b simultaneously. This minimization is the "least" in the least squares terminology and can be done by first substituting in for the Y_{calc} values in terms of m, b, and X_i, so that χ^2 is written as a function of m and b,

$$\chi^2 = \sum_{i=1}^{N} \left(mX_i + b - Y_i \right)^2 \tag{6}$$

and then differentiating χ^2 with respect to m and b separately:

$$\frac{\partial \chi^2}{\partial m} = \sum_{i=1}^{N} 2(mX_i + b - Y_i)X_i \tag{7}$$

$$\frac{\partial \chi^2}{\partial b} = \sum_{i=1}^{N} 2(mX_i + b - Y_i) \tag{8}$$

(Partial derivatives are needed for this. The partial derivative with respect to m is obtained by treating m as the variable and b and all the X_i and Y_i as constants, and applying the usual rules of differentiation. The partial derivative with respect to b is obtained by interchanging the roles of m and b, taking b as the variable and treating m as a constant and differentiating.) The minimum of χ^2 can be obtained by setting these two derivatives equal to zero and solving the resulting two equations simultaneously. The equations above (eqs. (7) and (8)), set equal to zero, can be expanded and cast into a slightly more orderly form before solving simultaneously:

$$m \sum_{i=1}^{N} X_i^2 + b \sum_{i=1}^{N} X_i - \sum_{i=1}^{N} X_i Y_i = 0 \tag{9}$$

$$m \sum_{i=1}^{N} X_i + Nb - \sum_{i=1}^{N} Y_i = 0 \tag{10}$$

These two equations may be solved for m and b in terms of the measured X_i and Y_i values. The solution is not difficult, but it may seem intimidating due to the summation notation. A bit of patience yields the results:

$$m = \frac{N \sum (X_i Y_i) - (\sum X_i)(\sum Y_i)}{N \sum (X_i^2) - (\sum X_i)^2} \tag{11}$$

$$b = \frac{\sum (X_i^2) \sum Y_i - \sum X_i \sum (X_i Y_i)}{N \sum (X_i^2) - (\sum X_i)^2} \tag{12}$$

These values of m and b depend only on the measured data and so can be calculated from the measurements. Note that the denominators for both are the same. For a small number of data points it's not unreasonable to do this with a hand calculator. Most scientific calculators have this as a standard function, usually called linear regression. This approach provides a systematic method of obtaining the best values for the slope and intercept, and therefore the physical parameters they represent. It automatically takes into account all the data collected. In evaluating the expressions in eqs. (11) and (12), it is worthwhile to evaluate each of the summations needed separately as an intermediate step, and then assemble the final results according to the prescriptions of eqs. (11) and (12). It's always smart to compare the results of this least squares calculation with values deduced by a quick "by eye" calculation from the graph or plot the line specified by the slope ad intercept on the graph with the data to see if it really "fits" the data. If there is disagreement the work should be double-checked. Remember to include units!

The method of least-squares can be applied to arbitrary functions for $Y_{calc}(X)$, not just to linear equations described here. This method can be used to evaluate many different physical models. More complicated functions can be much more difficult to use in the minimization of χ^2. In such cases the minimization is usually done numerically with the aid of a computer. The linear model presented here is especially convenient because the optimum or best values for the parameters (m and b) can be found in the closed form shown in eqs. (11) and (12).

Additional References:

Discussions of least squares analysis and curve fitting can be found in:

The Art of Experimental Physics, by D. W. Preston and E. R. Dietz, John Wiley and Sons, NY, 1991, especially pages 23-28.

A Practical Guide to Data Analysis for Physical Science Students by Louis Lyons, Cambridge University Press, Cambridge, 1991.

Appendix E
Resistor color code

Black	0
Brown	1
Red	2
Orange	3
Yellow	4
Green	5
Blue	6
Violet	7
Gray	8
White	9

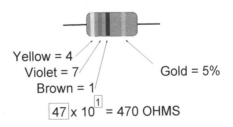

Yellow = 4
Violet = 7
Brown = 1
Gold = 5%

$47 \times 10^{1} = 470$ OHMS

For carbon film or carbon composition resistors (5 or 10% tolerance), three colored bands are used to indicate the resistance value followed with another band to represent the tolerance or probable error. If the colors represent three digits decoded by the above list as a-b-c, the resistance value is given by $ab \times 10^c$.

Examples:

Yellow, Violet, Brown translates into 4-7-1 $\Rightarrow$ $47 \times 10^1 = 470\,\Omega$.
Brown, Black, Red translates into 1-0-2 $\Rightarrow$ $10 \times 10^2 = 1000\,\Omega = 1\text{k}\Omega$
Green, Brown, Yellow translates into 5-1-4 $\Rightarrow$ $51 \times 10^4 = 510\,\text{k}\Omega$

Tolerances are represented by gold (5%) and silver (10%) fourth bands.

Special cases for low resistance values are rarely encountered. If the *third* band is gold, multiply the first two digits by 0.1. If the third band is silver, multiply by 0.01.)

Resistance values are frequently written in circuit diagrams in units of kΩ (10^3 ohms) or MΩ (10^6 ohms). In many diagrams the Ω is omitted and you may see a notation such as "51k" or "51K" next to the schematic symbol for a resistor indicating 51000 ohms.

As a quick gauge of the resistance, look at the third band:

Brown indicates 100's of Ω's (see the example above)
Red indicates kΩ's (see the example above)
Orange indicates 10's of kΩ's (red red orange = 22,000 =22 kΩ)
Yellow indicates 100's of kΩ's (blue gray yellow = $68 \times 10^4 = 680\,\text{k}\Omega$)
Green indicates MΩ's (orange orange green = $33 \times 10^5 = 3.3\,\text{M}\Omega$)

Metal film resistors, which are generally manufactured to 1% tolerances, require a third significant figure to specify the value to 1%. Then there are four colored bands to represent the value—three significant figures and the multiplier—followed by a fifth band to encode the tolerance. For example, yellow-white-white-red 4-9-9-2 = $49900\,\Omega = 49.9\,\text{k}\Omega$. The tolerance band on 1% resistors is brown (1), which can lead to occasional confusion about which end of the colors to begin with. Careful inspection usually reveals the tolerance band is spaced away from the four bands coding the value.